W9-DAU-730

The New Imperialism

David Harvey

OXFORD
UNIVERSITY PRESS

OXFORD
UNIVERSITY PRESS

Great Clarendon Street, Oxford OX2 6DP

Oxford University Press is a department of the University of Oxford.
It furthers the University's objective of excellence in research, scholarship,
and education by publishing worldwide in

Oxford New York

Auckland Bangkok Buenos Aires Cape Town Chennai
Dar es Salaam Delhi Hong Kong Istanbul Karachi Kolkata
Kuala Lumpur Madrid Melbourne Mexico City Mumbai Nairobi
São Paulo Shanghai Taipei Tokyo Toronto

Oxford is a registered trade mark of Oxford University Press
in the UK and in certain other countries

Published in the United States
by Oxford University Press Inc., New York

British Library Cataloguing in Publication Data
Data available

Library of Congress Cataloging in Publication Data
Data available

ISBN 0–19–926431–7

1 3 5 7 9 10 8 6 4 2

Typeset by Hope Services (Abingdon) Ltd.
Printed in Great Britain
on acid-free paper by
Biddles Ltd,
Guildford and King's Lynn

Contents

Preface

The Clarendon Lectures were delivered in the School of Geography and the Environment at Oxford University on 5, 6, and 7 February 2003. The timing is significant. War against Iraq, though seemingly imminent, had yet to begin and the faint hope still stirred that it could be stopped. That hope was bolstered by the huge global demonstrations, with a million or so people on the streets of London and Barcelona and impressive numbers recorded in many cities elsewhere throughout the world, including the United States, on 15 February. Sentiment within the Security Council of the United Nations largely supported the view that the threats posed by what everyone agreed was a barbaric and despotic regime could be resolved by diplomatic means. In spite of this opposition, military action against Iraq was initiated at the behest of the United States, supported most conspicuously by Britain and Spain, on 20 March. At the time of writing the outcome of the war, though not in doubt militarily, is still unclear. Will it end up being, or appearing to be, a colonial occupation, a US-imposed clientelist regime, or a genuine liberation?

Preface

On the one hand, these fast-moving events made it very difficult to devise a set of lectures on the topic of 'the new imperialism'. But, on the other hand, the very nature of these events and the threats they posed economically, politically, and militarily to global security made some sort of in-depth analysis imperative. I therefore determined to try as best I could to penetrate beneath the surface flux to divine some of the deeper currents in the making of the world's historical geography that might shed some light on why we have arrived at such a dangerous and difficult conjuncture.

In pursuance of that objective I gained much from sitting in on a year-long seminar organized on the topic of 'Imperialism' by Neil Smith and Omar Dahbour in the Center for Place, Culture and Politics at the CUNY Graduate Center. I wish to acknowledge the help of Neil, Omar, and the participants in that seminar in shaping many of my insights. Several colleagues in the Anthropology Program at CUNY likewise commented freely on my topic, and I thank Louise Lennihan, Don Robotham, Ida Susser, Jane Schneider, Talal Assad, and particularly Michael Blim and the students who participated in our joint seminar on 'Land, Labor, and Capital' for their input. The initial idea for some sort of intervention along the lines I here construct first vaguely occurred to me in a joint seminar I taught with Giovanni Arrighi at Johns Hopkins. I owe Giovanni a special debt. I am grateful to my colleagues in the Oxford School of Geography for the invitation to return to my old haunts and deliver these lectures at such an appropriate time and in such an appropriate place. I particularly want to thank Maria Kaika, Jack Langton, and Erik Swyngedouw for their warm welcome as well as for their intense interest in the topic. Anne Ashby of Oxford University Press proved most helpful and, as always, Jan Burke played her inestimable part in galvanizing me into action. Over the years I have gained much from interactions with others far too numerous to mention here. I hope I have put their individual and collective wisdom and understanding to good use in these lectures.

D.H.

1
All About Oil

My aim is to look at the current condition of global capitalism and the role that a 'new' imperialism might be playing within it. I do so from the perspective of the long durée and through the lens of what I call historical-geographical materialism. I seek to uncover some of the deeper transformations occurring beneath all the surface turbulence and volatility, and so open up a terrain of debate as to how we might best interpret and react to our present situation.

The longest durée any of us can actually experience is, of course, a lifetime. My first understandings of the world were formed during the Second World War and its immediate aftermath. Then, the idea of the British empire still had resonance and meaning. The world seemed open to me because so many spaces on the world map were coloured red, an empire upon which the sun never set. If I needed any additional proof of ownership, I could turn to my stamp collection—the head of the British monarch was on stamps from India, Sarawak, Rhodesia, Nyasaland, Nigeria, Ceylon, Jamaica . . . But I soon had to recognize that British power was in decline. The empire was crumbling at an alarming rate. Britain had ceded global power to the United States and the map of the world started to change colour as decolonization gathered pace. The traumatic events of Indian independence and partition in 1947 signalled the beginning of the end. At first I was given to understand that the trauma was a typical example of what happens when 'sensible' and 'fair' British rule gets replaced by irrational native passions and reversions to ancient prejudices (a framework for understanding the world that was and is not confined to Britain and has exhibited remarkable durability). But as struggles around decolonization became fiercer, so the seamier and more nefarious side of imperial rule became more salient. This culminated, for me and for many others of my generation, in the Anglo–French attempt to take back the Suez Canal in 1956. On that occasion it was the United States that rapped Britain and France over the knuckles for resorting to war to topple an Arab leader, Nasser, who, in Western eyes, was every bit as threatening and as 'evil' as Saddam Hussein is now depicted. Eisenhower preferred peaceful containment to war, and it is fair to say that the global reputation of the United States for leadership rose just as that of Britain and France fell precipitously. I found it hard after Suez to deny the perfidious side of a nakedly self-interested and rapidly fading but distinctively British imperialism.

Things looked very different to a young student from the Bronx who came to Oxford in the early 1960s. Marshall Berman records how he could not stand the 'languid young men who looked like extras from *Brideshead Revisited*, who slouched around in tuxedos (which often looked like they'd been slept in), vegetating while their fathers owned the British Empire and the world. Or at least they acted like their fathers owned the world. I knew how much of it really was an act: the Empire was *kaput*; the children of its ruling class were living on trust funds that were worth less every year, and inheriting companies that were going broke . . . at least I knew I was moving up in the world.'[1] I wonder how he feels now, with all those failed 'dot.com' companies littering the American landscape, accounting scandals, the catastrophic decline in stock markets that has destroyed a good chunk of everyone's pension rights, and sudden belligerent claims, most notably represented by the front cover of the *New York Times* Magazine for 5 January 2003: 'American Empire: Get Used to It.'[2] For me, it feels passing strange to come to consciousness of the world at the moment of one empire's passing and to come to retirement age at a moment of such public proclamations of the official birth of another.

Michael Ignatieff, the author of the *New York Times* piece, forcefully reiterates an earlier assertion (also in the *New York Times* Magazine of 28 July 2002) that 'America's entire war on terror is an exercise in imperialism. This may come as a shock to Americans, who don't like to think of their country as an empire. But what else can you call America's legions of soldiers, spooks and special forces straddling the globe?' The US can no longer favour empire 'lite' or expect to do it on the cheap, he argues. It should be prepared to take on a more serious and more permanent role, be prepared to stay for the long term to realize major transformative objectives. That such a mainstream publication should give such prominence to the idea of American empire has significance. And

Ignatieff is not alone in these assertions. Max Boot, an editor of the *Wall Street Journal*, opines that 'a dose of U.S. imperialism may be the best response to terrorism'. America must be more expansive, he says: 'Afghanistan and other troubled lands today cry out for the sort of enlightened foreign administration once provided by self-confident Englishmen in jodhpurs and pith helmets'. With their grand imperial traditions so nostalgically depicted, the British also got in on the act. The conservative historian Niall Ferguson (whose TV series and accompanying book document, in true patriotic fashion, not only the heroic deeds of Britain's empire-builders but also the peace, prosperity, and well-being that this empire supposedly gave to the world) advises that the US must stiffen its resolve, shell out the money, and 'make the transition from informal to formal empire'. A 'new imperialism', many now assert, is already in operation, but it calls for more explicit acknowledgement and a more solid commitment if it is to establish a Pax Americana that can bestow the same benefits upon the world as the Pax Brittanica secured in the last half of the nineteenth century.[3]

This is a commitment that President Bush seems willing to make in spite of his declaration in a West Point speech that 'America has no empire to extend or utopia to establish'. 9/11, he wrote in an op-ed piece for the *New York Times* on the anniversary of that tragedy, has clarified America's role in the world and opened up great opportunities. 'We will use our position of unparalleled strength and influence to build an atmosphere of international order and openness in which progress and liberty can flourish in many nations. A peaceful world of growing freedom serves American long-term interests, reflects enduring American ideals and unites America's allies. . . . We seek a just peace', he wrote while preparing to go to war, 'where repression, resentment and poverty are replaced with the hope of democracy, development, free markets and free trade', these last two having 'proved their ability to lift whole societies out of poverty'. The United States, he asserted, 'will promote moderation, tolerance and the nonnegotiable demands of human dignity—the rule of law, limits on the power of the state, and respect for women, private property, free speech and equal justice'. Today, he concluded, 'humanity holds in its hands the opportunity to offer freedom's triumph over all its age-old foes. The United States welcomes its responsibility to lead in this great mission.' This same language appeared in the prologue to the National Defense Strategy document issued shortly thereafter.[4] This may not amount to a formal declaration of empire but it most certainly is a declaration redolent of imperial intent.

There have been many different kinds of empire (Roman, Ottoman, Imperial Chinese, Russian, Soviet, Austro-Hungarian, Napoleonic, British, French, etc.). From this motley crew we can easily conclude there is considerable room for manoeuvre as to how empire should be construed, administered, and actively constructed. Different and sometimes rival conceptions of empire can even become internalized in the same space. Imperial China went through a strong expansionary phase of oceanic exploration only to suddenly and mysteriously withdraw into itself. American imperialism since the Second World War has lurched in unstable fashion from one vague (because always left undiscussed) conception of empire to another. If Bush the younger betrays a certain Napoleonic impulse, wanting to march on Baghdad and perhaps afterwards on Tehran (where some of the hawks in the administration apparently believe 'real men' truly belong), Clinton's approach (interestingly dubbed 'effeminate' by the Bush administration) more resembled that of the Ottoman empire at its height. Highly centralized within the US Treasury, where Rubin and then Summers were commanding figures, soft power was preferred to hard, and the rest of the world was treated with considerable multicultural tolerance. Politics was conducted in multilateral rather than unilateral terms. The construction of American imperial power under Roosevelt, Truman, and Eisenhower right through to Nixon, on the other hand, mirrored the subordinate client state approach of the Soviets rather than anything else, with the difference that Japan, unlike Hungary or Poland, was left free to develop its own economy provided it remained politically and militarily compliant with US wishes. The actually existing American empire was acquired, Ignatieff suggests, not in a fit of absent-mindedness (as the British liked to claim), but in a state of denial: imperial actions on the part of the United States were not to be talked of as such, nor were they allowed to have any ramifications for the domestic situation. It was this that produced 'empire lite' rather than an empire of solid, long-term commitment.[5]

There are plenty of people on what might be called the 'traditional left' who hold that the US has been an imperial power for at least a century or more. Fulsome analyses of American imperialism were available in the 1960s, particularly focusing on the US role in Latin America and

South-East Asia. There were substantive disputes between the then newly minted dependency theorists (like Frank) and those more inclined to take Hobson, Hilferding, Lenin, Luxemburg, and other turn-of-the-century theorists at their word. And Mao certainly considered US imperialism the primary contradiction with which he had to contend. But the publication of Hardt and Negri's *Empire* in 2000, and the controversy that surrounded it, challenged traditional debates and suggested that left opposition had to be rethought in relation to a decentred configuration of empire that had many new, postmodern, qualities. While critical of this line of argument, many others on the left began to recognize that the forces of globalization (however those might be construed) were creating a novel situation that required a new framework of analysis.[6] The overt recognition of empire and of imperialism by those on the right as well as those of a liberal persuasion was therefore a welcome acknowledgement of what had long been the case. But it also indicated that imperialism might now be taking on a rather different allure. The effect has been to turn the questions of empire and of imperialism into open topics of debate across the political spectrum (it was noteworthy that Hardt and Negri's work gained attention in the mainstream media). But this then poses the further question: what, if anything, is new about all this?

I approach this question in the first instance through an examination of contemporary events. The US, backed by Britain, Spain, and Australia and with the approval of several other states, has gone to war with Iraq. But it has done so in the midst of fierce opposition from several traditional allies, most notably France and Germany, as well as from long-standing opponents, most notably Russia and China. Popular mobilizations against the war have occurred around the world and there is a sense of bewilderment on the part of many as to why the Bush administration became fixated upon such a course of action. The evidence suggests that there is something deep at work in this. But it is hard to see what it is. These deeper meanings have to be excavated from beneath an incredible surface froth of misleading rhetoric and disinformation.

A Tale of Two Oil Producers

The coup that overthrew President Chavez of Venezuela in April 2002 was greeted with euphoria in Washington. The new president—a businessman—was instantly recognized and the hope expressed that stability and order would return to the country, thus creating the basis for solid future development. The *New York Times* editorialized in identical language. Most people in Latin America, however, immediately saw the hand of the CIA and recalled what the Chileans now ironically refer to as 'their little September 11th' of 1973 when the democratically elected socialist, Salvador Allende, was overthrown in a brutal coup by General Augusto Pinochet. In the State Department archive of that event there is a CIA cable that reads 'It is firm and continuing policy that Allende be overthrown by a coup . . . We are to continue to generate maximum pressures toward this end utilizing every appropriate resource. It is imperative that these actions be implemented clandestinely and securely so that United States Government and American hand be well hidden.'[7]

It is not hard to imagine similar cables with respect to Venezuela adorning the State Department website at some future date.

The coup was reversed three days later and Chavez then came back to power. The State Department soberly denied any prior knowledge about anything, saying it was all an internal matter. It was to be hoped that a peaceful, democratic, and constitutional solution to the difficulties would be arrived at, they said. The *New York Times* editorial followed suit, merely adding that perhaps it was not a good idea to embrace the overthrow of a democratically elected regime, however obnoxious, too readily if one of America's fundamental values was support for democracy.

The parallel with Iraq, incidentally another key member of OPEC, is instructive. There, the United States claims to have an interest in establishing democracy. Of course it had earlier overthrown the democratically elected Mossadegh of Iran in 1953 and installed the dictatorial Shah of Iran upon the throne. So presumably it is only democratically elected governments of a certain sort that will be tolerated. But in this instance the claim to want to democratize Iraq and the whole region was but one claim among a welter of often conflicting explanations given as to why it was important to be prepared to go to war. Most people, even supporters, were perplexed and confused by the rationalizations. It proved hard to get behind the clutter of disinformation and the perpetually shifting arguments. An early attempt to connect Iraq to the anthrax attacks in the United States failed miserably. While Iraq has a ghastly record of using biological and chemical weapons, most of this occurred when the United

States was supporting Iraq against Iran, and the State Department deliberately misled the world into thinking that both sides were then resorting to such heinous methods when it knew full well that Iraq was the sole offender.[8] The equally ghastly record on human rights warrants consideration, but this hardly makes sense as policy when the US government proffers military assistance to Algeria—a country that vies with Iraq in terms of violent human rights abuses to suppress its Islamicist opposition (120,000 deaths estimated in the last eight years). William Burns, US Assistant Secretary of State for the Middle East, even went so far as to say that 'we have much to learn from the Algerians when it comes to controlling terror'.[9] This may explain why the issue of when torture might be justified suddenly became a matter of public debate in the United States (again given prominence in the *New York Times*).

Then there is the problem of the weapons of mass destruction. What Iraq does have is hard to know, but its military capacity was so degraded during and after the Gulf War that even CIA assessments considered it to be no real threat to the peace of the region. This made assertions that Iraq was a threat to the United States (with President Bush bizarrely going so far as to assert that an Iraqi attack upon the United States would do great damage to the US economy) sound foolish. The CIA concluded that Saddam would use biological and chemical weapons, if he had them, only if provoked. This made it doubly hard to explain why it was that the US seemed so determined to provoke him. Most probably Iraq is trying to go nuclear, but then so are a lot of other countries, with North Korea openly declaring so. The weapons inspectors, when finally allowed in, could not find that much. In any case, regime change was the original objective and disarmament only became prominent as a reason to invoke the authority of the United Nations, given that the UN Charter does not allow for pre-emptive attacks. And if all that failed, then Saddam had to go because he was a liar (an appellation that sticks to so many politicians that it quickly became a joke), ruthless (but then so is Sharon), reckless (not proven), or an incarnation of evil that had to be combated as if war in the Middle East was an episode in some long-running medieval morality play (with Saddam cast as Mordor and George Bush as the brave Frodo accompanied by Blair as his faithful Sam). In the end it was all made to sound as if the US and Britain had become committed to some high-sounding moral mission to free the Iraqi people no matter what and implant American-style enlightenment in the Middle East.

In all of this, it was hard not to have the impression that something very important was being concealed behind a whole series of smokescreens. At first it seemed plausible that there was secret information that could not be revealed, but every time there was an attempt to reveal something from the secret archive it appeared either trivial, easily refuted, or, in the case of the British revelations that were plagiarized without acknowledgement from a five-year-old doctoral dissertation (part of which had already been published in *Foreign Affairs*), was so sloppily researched as to be hard to take seriously. Leaks from the intelligence community in the United States suggested that some of its members were unhappy with the way in which their information was being doctored by the administration. Small wonder that the balance of world opinion, in spite of a bellicose press (all 175 newspapers owned by Murdoch world-wide, staffed by editors supposedly chosen for their independence, unanimously proclaimed war was a good thing, as did various others owned by media tycoons), and a lot of hectoring from the politicians, remained profoundly sceptical, if not outright opposed to war.

So what is really going on? The stated reasons fail to convince; they simply do not add up to a compelling case. What, then, might the unstated reasons be? And here we may have to confront the fact that these reasons may not even be well understood by the principal actors in the drama, or, if they are understood, that they are being actively suppressed or denied.

The Inner Dialectic of US Civil Society

Shortly before the German elections in 2002, the German Minister of Justice caused a furore by suggesting that the adventurism of the Bush administration abroad was designed to divert attention from its difficulties at home. Her mistake was to add that this had been one of Hitler's tactics too, and for that she had to go. The effect, unfortunately, was to bury any serious discussion on the first part of her proposition.

There is indeed a long history of governments in trouble domestically seeking to solve their problems either by foreign adventures or by manufacturing foreign threats to consolidate solidarities at home. The idea warrants serious consideration in this case because the internal condition of the United States during 2002 was in many respects more parlous than it had been for many years. The recession that began early in 2001 (and which was prodded onwards by the shock of 9/11) would not go away. Unemployment was rising and the sense of economic insecurity was palpable. Corporate scandals cascaded over each other and seemingly solid corporate empires were literally dissolving overnight. Accounting failures (as well as outright corruption) and failures of regulation were bringing Wall Street into disrepute, and stocks and other asset values were plunging. Pension funds lost between a quarter and a third of their value (if they did not totally disappear, as in the case of the funds of Enron employees), and the retirement prospects of the middle class took a serious hit. Health care was in a mess, federal, state, and local government surpluses were evaporating fast, and deficits began to loom larger and larger. The current account balance with the rest of the world was going from bad to worse as the United States became the biggest debtor nation of all time. Social inequality had long been on the increase but the tax-cut fetish of the administration seemed set fair to increase it further. Environmental protections were being gutted, and there was a deep reluctance to reimpose any regulatory framework on the markets even in the face of clear evidence of market failure. To top it all, the president had been elected by a five-to-four vote of the Supreme Court rather than by the people. His legitimacy was questioned by at least half the population on the eve of 9/11. The only thing to prevent the political annihilation of the Republicans was the intense solidarity—verging on a nationalist revival—created around the events of 9/11 and the anthrax scare (still,

curiously, not solved and largely forgotten except as a har-
binger of the sort of thing Saddam would be only too
ready to inflict). While Afghanistan submitted to US
power quickly and (for the Americans) bloodlessly, Osama
had not been found 'dead or alive' and the war on terrorism was not producing very much in the way of spectacular results. What better moment, then, than to switch the
focus to Iraq, as one of the key pillars in 'an axis of evil'
that the hawkish members of the Bush administration had
wanted to go after militarily ever since the inconclusive
end to the Gulf War? That the diversionary tactic worked,
at least in the short run, is a matter of history. The
American public by and large accepted the idea that there
was some sort of connection between al Qaeda and
Saddam's regime and that the latter was in any case a
sufficiently dangerous and evil enemy as to warrant military action to remove him. And en route the Republicans
were able to consolidate political power through the
Congressional elections, and the president could shed the
air of illegitimacy that had hung over his election.

But there may be something far deeper at work here
that converts what looks like shallow political oppor-
tunism into a compelling and enduring political force
within the geopolitical history of the United States. To
begin with, fear of Iraqi power and of a potentially disruptive pan-Arabic movement had long lurked within
successive US administrations. Colin Powell had laid
military contingency plans to deal with Iraq prior to the
first Gulf War. Paul Wolfowitz, who became Bush's
Deputy Secretary of Defense, had explicitly argued for
regime change in Iraq as early as 1992 and publicly proclaimed so throughout the 1990s. Regime change became

accepted policy in the Clinton administration. A neo-conservative group brought together under the rubric of the *Project for the New American Century* in 1997 insisted on this as a key objective and urged that it be done militarily. The group included Rumsfeld, Wolfowitz, Armitage, Perle, and several others who were to form the core of Bush's defence and foreign policy team. Geo-strategically, then, Iraq had long been in the sights of this group. But they recognized in a 1999 report that it would take 'a catastrophic and catalyzing event, like a new Pearl Harbor' to make a military strike acceptable internation-ally and domestically. 9/11 provided the opportunity, if only they could make a connection between Saddam and al Qaeda.[10] With most of the American public uncaring and uninformed about almost anything geographical, it proved fairly easy to parlay the hunt for terrorists into a campaign to hunt down and remove Saddam. The rest of the world was not so convinced.

There is yet another dimension to this internal dynamic that needs to be understood. The US is a quite extraordinary multicultural immigrant society driven by a fierce competitive individualism that perpetually revolu-tionizes social, economic, and political life. These forces render democracy chronically unstable, difficult if not impossible to command except through the corruption of financial power. There are times when the whole country appears so unruly as to be ungovernable. Hannah Arendt captures what such a civil society is about exactly:

Since power is essentially only a means to an end a community based solely on power must decay in the calm of order and sta-bility; its complete security reveals that it is built on sand. Only by acquiring more power can it guarantee the status quo; only

15

by constantly extending its authority and only through process of power accumulation can it remain stable. Hobbes's Commonwealth is a vacillating structure and must always provide itself with new props from outside; otherwise it would collapse overnight into the aimless, senseless chaos of the private interests from which it sprang. . . . [The] ever-present possibility of war guarantees the Commonwealth a prospect of permanence because it makes it possible for the state to increase its power at the expense of other states.[11]

The Cold War was over and the threat of Russians with snow on their boots plodding down across Canada was no longer credible. During the 1990s there was no clear enemy and the booming economy within the United States should have guaranteed an unparalleled level of contentment and satisfaction throughout all but the most underprivileged and marginalized elements in civil society. Yet, as Arendt might have predicted, the 1990s turned out to be one of the most unpleasant decades in US history. Competition was vicious, the avatars of the 'new economy' became millionaires overnight and flaunted their wealth, scams and fraudulent schemes proliferated, scandals (both real and imagined) were everywhere embraced with gusto, vicious rumours circulated about assassinations plotted in the White House, an attempt was made to impeach the president, talk-show hosts Howard Stern and Rush Limbaugh typified a media totally out of control, Los Angeles erupted in riots, Waco and Oklahoma symbolized a penchant for internal opposition and violence that had long remained latent, teenagers shot and killed their classmates in Columbine, irrational exuberance prevailed over common sense, and corporate corruption of the political process was blatant. Civil society was, in short, far from civil. Society seemed to be fragmenting and flying apart at an alarming rate. It seemed, as Arendt would put it, in the process of collapsing back into the aimless, senseless chaos of private interests.

Part of George Bush's electoral appeal in 2000, I suspect, was his promise of providing a strong-minded and tough moral compass to a civil society spiralling out of control. All of his key appointments came from the ranks of neo-conservatives with a bent, like John Ashcroft as Attorney General, for authoritarian state action. Neo-conservatism displaced neo-liberalism of the sort that Clinton had championed. But it was, of course, 9/11 that provided the impetus to break with the dissolute ways of the 1990s. It provided the political opening not only to assert a national purpose and to proclaim national solidarity, but also to impose order and stability on civil society at home. It was the war on terror, swiftly followed by the prospect of war with Iraq, which allowed the state to accumulate more power. The engagement with Iraq was far more than a mere diversion from difficulties at home; it was a grand opportunity to impose a new sense of social order at home and bring the commonwealth to heel. Criticism was silenced as unpatriotic. The evil enemy without became the prime force through which to exorcise or tame the devils lurking within. This relation between the internal and external conditions of political power has played a significant if largely hidden role in the dynamics that have fuelled the conflict with Iraq. We will have occasion to return to it more than once in what follows.

All About Oil

Opponents of war with Iraq frequently depict the conflict as all about oil. The US government either dismisses that claim out of hand as preposterous or ignores the question entirely. There is no question that oil is crucial. But exactly how and in what sense is not so easy to determine.

A narrow conspiracy thesis rests on the idea that the government in Washington is nothing more than an oil mafia that has usurped the public domain. This idea is supported by the close connections of Bush and Cheney to oil interests, coupled with reports that Halliburton, Vice-President Cheney's old company, stands to gain nearly a billion dollars in contracts for oil services in the immediate aftermath of the war.[12] While none of this hurts, I cannot imagine that the political-military establishment as a whole or corporate interests in general would countenance war on such grounds. It is of course the case that US and British oil companies had been excluded from Iraq and that French, Russian, and Chinese companies have been favoured. The opposition to war as opposed to peaceful disarmament had been articulated most strongly by those countries that already had concessions. If disarmament was certified then UN sanctions would have been lifted and the existing concessionaires would have benefited. Regime change through war means concessions will almost certainly be renegotiated. But Iraq owns the oil, and the prospects for the oil companies even after regime change are not necessarily so rosy. The only scenario that would work would be if some post-war US administration took over the Iraqi oil company or set up some front organization—such as an

18

international consortium in which the US, as in the IMF, would have veto power—to manage the exploitation and use of the oil. But all of this would be very difficult to negotiate without stirring up strong antagonisms both within Iraq and between capitalist powers.

There is, however, an even grander perspective from which to understand the oil question. It can be captured in the following proposition: whoever controls the Middle East controls the global oil spigot and whoever controls the global oil spigot can control the global economy, at least for the near future.[13]

We should not, therefore, think solely of Iraq, but consider the geopolitical condition and significance of the Middle East as a whole in relation to global capitalism. And this point is made in the official rhetoric. The plan for regime change in Iraq overtly states that the influence of a democratic and pro-US government would be beneficial throughout the whole region, and perhaps even influence similar regime changes elsewhere (Iran and Syria being the most obvious targets, with Saudi Arabia not far behind). There are even those in the administration hubristic enough to think that a general conflagration in the region would provide an opportunity to redraw the whole map of the Middle East (much as happened in the old Soviet Union and Yugoslavia). State formation in the region, after all, had largely occurred as a side-bar to the Versailles settlement after the First World War. This settlement is generally acknowledged to have betrayed Arab interests and imposed a configuration of states reflecting British and French imperial interests. This configuration could be viewed as anachronistic and dysfunctional. A comprehensive settlement might cater to some separatist interests (federal status for the Kurds within Iraq, for example, and perhaps the break-up of Iraq into a southern Shi'ite state based on Basra). Most important of all, it might permit a settlement of the Israeli–Palestinian question through formation of a greater Palestinian state incorporating Jordan and perhaps part of Saudi Arabia. Against this there are very strong opinions in the UN that preservation of the territorial integrity of Iraq as it exists now must be a primary objective in any post-war settlement, and to this the United States has at least nominally agreed.

The US has a long-standing geopolitical interest in the region. Crucial to the whole concept of global control as worked out during the Second World War was

control of the Middle East, which was regarded as part of the old British Empire, and absolutely essential for the economic, military, and political control of the globe—not least of all because it was the repository of most of the world's proven oil reserves. The United States thus began a long series of overt and covert operations in the region in the 1950s, the foremost of which was the 1953 overthrow of the democratically elected Mossadegh government in Iran, which had nationalized foreign-owned oil companies. The success of the US drive was clear. Between 1940 and 1967, US companies increased their control of Middle Eastern oil reserves from 10 percent to close to 60 percent while reserves under British control decreased from 72 percent in 1940 to 30 percent in 1967.[14]

In the late 1960s the British abandoned any military presence east of Suez, leaving the US in sole command. Because of Vietnam, the US chose to use the surrogate states of Iran and Saudi Arabia to look after its proliferating interests in the region. It also looked to its particularly strong and almost unquestioning support of Israel to create there a solid outpost of American surrogate power in the region. But first the oil boycott and price hike of 1973 organized through OPEC, and then the fall of the Shah of Iran in 1979, made this solution of indirect rule through distant surrogates untenable. President Carter enunciated the doctrine that the United States would not under any circumstances allow an interruption of the flow of Gulf oil. This meant a commitment to keeping the Strait of Hormuz open (for the delivery and distribution systems are every bit as important as the oilfields themselves) and a permanent military presence in the region, plus the formation of a Rapid Deployment Force to deal with any emergencies. The US covertly encouraged and supported Iraq's brutal and deadly war with Iran, but Iraq's growing power sparked planning (initiated by Colin Powell) for a conflict with Iraq well before the Kuwait invasion occurred. Why the US ambassador to Iraq signalled that the US would not respond militarily to any Iraqi move into Kuwait is still a matter of controversy, with entrapment rather than simple though catastrophic misunderstanding one possible explanation.

The Gulf War, though inconclusive with respect to Iraq, brought a much stronger US military presence in the region. This continued unabated during the Clinton administration. Joint patrols of the 'no-fly zones' with the British entailed a continuous low-level aerial combat and missile attacks on Iraqi military facilities. Joseph Nye, an official in the Clinton administration and generally an advocate of 'soft power', nevertheless categorically stated that the US would not hesitate to use military force in the Gulf region and would do so unilaterally if necessary, if

US interests were in any way threatened.[15] It took a strong build-up of US forces in 1997–8 to force the first round of weapons inspectors into Iraq to certify that the terms of the peace agreement on Iraq's disarmament were being observed. Missile attacks and aerial conflict escalated. To support its efforts, the US set up the Gulf Cooperation Council with Saudi Arabia, Kuwait, and other states, selling them military equipment as a back-up for US forces in the region (a net $42 billion arms transfer—$23 billion to Saudi Arabia alone—occurred during the 1990s). US forces were being pre-positioned in the region during the 1990s and large stores of military equipment were established in Kuwait, Qatar, and Saudi Arabia giving the US an immediate ability to move. Military planning, with the Cold War over, shifted to being able to fight two regional wars at once, and Iraq and North Korea were chosen as a planning exercise. By the late 1990s, more than 20,000 military personnel were deployed in the region at an annual cost of $4–5 billion a year.

I briefly review this history here in order to make two basic points. Since 1945 there has been a steady escalation of US involvement in the region, marked by a significant break after 1980 as the involvement came to depend more and more on a direct military presence. Secondly, the conflict with Iraq is of long standing, and planning for some sort of military denouement was in the works even before the last Gulf War started. The only difference between the Clinton years and now is that the mask has come off and bellicosity has displaced a certain reticence, in part because of the post-9/11 atmosphere within the United States that makes overt and unilateral military action more politically acceptable. Viewed geopolitically and in the long term, some confrontation with Iraq appeared inevitable unless it became a client state of the US, like Saudi Arabia. But why this geopolitical thrust? Again, the answer has everything to do with oil.

At any one time, the status of global oil reserves is a matter of conjecture. Oil companies are notoriously reticent to say what they know and on occasion deliberately mislead. Estimates of reserves often differ wildly. Most accounts suggest, however, that the rate of exploitation of oil reserves has exceeded the rate of discovery since 1980 or so. Oil is slowly becoming increasingly scarce. We do know that many fields are past their peak and that within a decade or so many of the world's present oilfields will be depleted. This is the case for domestic US, North Sea, Canadian, Russian, and (more ominously) Chinese production. While other oilfields have a longer life, the only fields that look set to last fifty years or more are those in Iran, Iraq, Saudi Arabia, the United Arab Emirates, and Kuwait. While new discoveries could change this picture, most strategic thinkers have to confront the increasing significance of the Middle East as the key provider of oil over time. On the demand side we see that the United States is increasingly dependent upon foreign imports, that the dynamic centres of economic growth in East and South-East Asia are almost bereft of significant oil reserves (with demand in China now escalating at a phenomenal rate), and that Europe (with the exception of Britain and Norway) is likewise totally dependent on imported oil. Alternatives to oil are being explored, but there is very little chance that these will be serious contenders (given the barriers erected by the oil companies and other vested interests) for several decades. Access to

Middle Eastern oil is now, therefore, a crucial security issue for the United States, as it is for the global economy as a whole.

This immediately poses the problem of US motivation in seeking tighter military and strategic control, unilaterally if necessary. Thomas Friedman argues, for example, that 'there is nothing illegitimate or immoral about the US being concerned that an evil, megalomaniacal dictator might acquire excessive influence over the natural resource that powers the world's industrial base'. But we have to be careful to convey to the public and reassure the world that the intention is 'to protect the world's right to economic survival' rather than our own right to indulge ourselves, that the US is 'acting for the benefit of the planet, not simply to fuel American excesses. . . . If we occupy Iraq and simply install a more pro-US autocrat to run the Iraqi gas station (as we have in other Arab oil states), then this war partly for oil would be immoral.'[16] Is the US, in short, exercising leadership and seeking to regulate the use of Middle Eastern oil in everyone's interests through consent? Or is it seeking domination to realize its own far narrower strategic interests? Friedman wishes to believe the former. But what if it is the latter?

If the US successfully engineers the overthrow of both Chavez and Saddam, if it can stabilize or reform an armed-to-the-teeth Saudi regime that is currently based on the shifting sands of authoritarian rule (and in imminent danger of falling into the hands of radicalized Islam), if it can move on (as seems it will likely seek to do) from Iraq to Iran and consolidate a strategic military presence in the central Asian republics and so dominate Caspian Basin oil reserves, then it might, through firm control of the global oil spigot, hope to keep effective control over the global economy for the next fifty years. Europe and Japan, as well as East and South-East Asia (now crucially including China) are heavily dependent on Gulf oil, and these are regional configurations of political-economic power that now pose a challenge to US global hegemony in the worlds of production and finance. What better way for the United States to ward off that competition and secure its own hegemonic position than to control the price, conditions, and distribution of the key economic resource upon which those competitors rely? And what better way to do that than to use the one line of force where the US still remains all-powerful—military might? There is also a military aspect to this argument. The military runs on oil. North Korea may have a sophisticated airforce, but it cannot use it much for lack of fuel. Not only does the US need to ensure its own military supplies, but any future military conflict with, say, China will be lopsided if the US has the power to cut off the oil flow to its opponent. But such lines of argument only make sense if the US has reason to fear that its dominant position within global capitalism is somehow threatened. It is to the economic rather than the military dimension to this question that I turn in Chapter 2 of this enquiry.

2

How America's Power Grew

Imperialism is a word that trips easily off the tongue. But it has such different meanings that it is difficult to use it without clarification as an analytic rather than a polemical term. I here define that special brand of it called 'capitalist imperialism' as a contradictory fusion of 'the politics of state and empire' (imperialism as a distinctively political project on the part of actors whose power is based in command of a territory and a capacity to mobilize its human and natural resources towards political, economic, and military ends) and 'the molecular processes of capital accumulation in space and time' (imperialism as a diffuse political-economic process in space and time in which command over and use of capital takes primacy). With the former I want to stress the political, diplomatic, and military strategies invoked and used by a state (or some collection of states operating as a political power bloc) as it struggles to assert its interests and achieve its goals in the world at large. With the latter, I focus on the ways in which economic power flows across and through continuous space, towards or away from territorial entities (such as states or regional power blocs) through the daily

26

practices of production, trade, commerce, capital flows, money transfers, labour migration, technology transfer, currency speculation, flows of information, cultural impulses, and the like.

What Arrighi refers to as the 'territorial' and the 'capitalist' logics of power are rather different from each other.[1] To begin with, the motivations and interests of agents differ. The capitalist holding money capital will wish to put it wherever profits can be had, and typically seeks to accumulate more capital. Politicians and statesmen typically seek outcomes that sustain or augment the power of their own state vis-à-vis other states. The capitalist seeks individual advantage and (though usually constrained by law) is responsible to no one other than his or her immediate social circle, while the statesman seeks a collective advantage and is constrained by the political and military situation of the state and is in some sense or other responsible to a citizenry or, more often, to an elite group, a class, a kinship structure, or some other social group. The capitalist operates in continuous space and time, whereas the politician operates in a territorialized space and, at least in democracies, in a temporality dictated by an electoral cycle. On the other hand, capitalist firms come and go, shift locations, merge, or go out of business, but states are long-lived entities, cannot migrate, and are, except under exceptional circumstances of geographical conquest, confined within fixed territorial boundaries.

The two logics contrast in other ways. Though the degree and modalities of public involvement vary greatly, the politics of state and empire of the sort we now experience are open to discussion and debate. Specific decisions have to be taken, such as whether or not to go to war with

Iraq, whether or not to do it unilaterally, how to deal with post-war difficulties, and the like. Foreign policy establishments and political/military experts debate these issues, and it would be rare indeed if there were no dissent. But clear decisions with all manner of ramifications have to be made. Strategic decisions of sometimes immense import (and not a few sometimes startling unintended consequences) are arrived at and implemented in the rough and tumble of the political process where variegated interests and opinions clash (sometimes even hinging on the particular beliefs or charisma of those in power or the outcome of personality conflicts between influential players).

The geographical processes of capital accumulation, on the other hand, are much more diffuse and less amenable to explicit political decision-making in this way. Individual (usually business, financial, and corporate) agency is everywhere at work and the molecular form makes for multiple forces that bump into each other, sometimes counteracting and at other times reinforcing certain aggregate trends. It is hard to manage these processes except indirectly, and then often only after the fact of already established trends. The institutional arrangements embedded within the state have, as we shall see, an influential role to play in setting the stage for capital accumulation. And there are monetary and fiscal levers and strings (of the sort that Alan Greenspan wields as Chairman of the Federal Reserve) as well as a range of fiscal and monetary modes of intervention (including taxation arrangements, redistributive policies, state provision of public goods, and direct planning) that clearly position the state as a powerful economic agent in its own right. But even in authoritarian states or those states dubbed 'developmental' by virtue of their strong inner connections between state policies, finance, and industrial development, we find the molecular processes often escape control. If I decide to buy a Toyota rather than a Ford, or see a Hollywood as opposed to a Bollywood movie, what does this do to the US balance of payments? If I transfer money from New York to needy relatives in Lebanon or Mexico what does this do to the financial balances between nations? It seems impossible to anticipate, and difficult even to keep track of the flows of capital and of money through the vagaries of the credit system. All sorts of psychological intangibles, such as investor or consumer confidence, enter into the picture as determinant forces. Thus did Keynes (drawing secretly on Marx) invoke 'the animal spirits' of the entrepreneur and the expectations of the financiers as crucial to the vigour and viability of capitalism. The best we can do is to anxiously monitor the data after the event, in the hope we can spot trends, second-guess what the market will do next, and apply some corrective to keep the system in a reasonably stable condition.

The fundamental point is to see the territorial and the capitalist logics of power as distinct from each other. Yet it is also undeniable that the two logics intertwine in complex and sometimes contradictory ways. The literature on imperialism and empire too often assumes an easy accord between them: that political-economic processes are guided by the strategies of state and empire and that states and empires always operate out of capitalistic motivations. In practice the two logics frequently tug against each other, sometimes to the point of outright antagonism. It would be hard to make sense of the Vietnam War or the invasion of Iraq, for example, solely in terms of the immediate requirements of capital accumulation. Indeed, a plausible case can be made that such ventures inhibit rather than enhance the fortunes of capital. But, by the same token, it is hard to make sense of the general territorial strategy of containment of Soviet Power by the United States after the Second World War—the strategy that set the stage for US intervention in Vietnam—without recognizing the compelling need felt on the part of business interests in the United States to keep as much of the world as possible open to capital accumulation through the expansion of trade, commerce, and opportunities for foreign investment. The relation between these two logics should be seen, therefore, as problematic and often contradictory (that is, dialectical) rather than as functional or one-sided. This dialectical relation sets the stage for an analysis of capitalist imperialism in terms of the intersection of these two distinctive but intertwined logics of power. The difficulty for concrete analyses of actual situations is to keep the two sides of this dialectic simultaneously in motion and not to lapse into either a solely political or a predominantly economic mode of argumentation.

It is not always easy to determine the relative importance of these two logics in generating social and political change. Was the USSR brought down by the strategic decision of the Reagan administration to launch an immense arms race and break the back of its economy? Or was it brought down by molecular changes within the body politic of the Soviet system (including, for example, the corrosive influence of money power or of capitalist cultural forms clandestinely entering from outside)? Are we now witnessing overt political claims about empire and the imperialism that goes with it within the United States at the political and territorial level, at the very moment when the flows of economic power and even cultural and moral influence are ebbing away from its shores into more diffuse regional power blocs (centred on Asia and Europe, for example)? Are we seeing the disintegration of US hegemony within the global system and the rise of a 'new regionalism' in political-economic power even as we see the United States acting as if it is the sole superpower to be obeyed? What dangers does this regionalization portend, given that the last period in which it dominated was the 1930s and that this collapsed under economic and political pressures into global war? Does the US have the power to reverse or control such regional fragmentation? These are the big questions that I will seek to address.

I will focus more closely on exactly how the molecular processes of capital accumulation work in Chapter 3. But I need to say something about them here in order to specify more clearly the constraints within which the territorial logic of power works. Imperialistic practices, from the perspective of capitalistic logic, are typically about exploiting the uneven geographical conditions under which capital accumulation occurs and also taking advantage of what I call the 'asymmetries' that inevitably arise out of spatial exchange relations. The latter get expressed through unfair and unequal exchange, spatially articulated monopoly powers, extortionate practices attached to restricted capital flows, and the extraction of monopoly rents. The equality condition usually presumed in perfectly functioning markets is violated, and the inequalities that result take on a specific spatial and geographical expression. The wealth and well-being of particular territories are augmented at the expense of others. Uneven geographical conditions do not merely arise out of the uneven patterning of natural resource endowments and locational advantages, but, even more importantly, are produced by the uneven ways in which wealth and power themselves become highly concentrated in certain places by virtue of asymmetrical exchange relations. It is here that the political dimension re-enters the picture. One of the state's key tasks is to try to preserve that pattern of asymmetries in exchange over space that works to its own advantage. If, for example, the US forces open capital markets around the world through the operations of the IMF (International Monetary Fund) and the WTO (World Trade Organization), it is because specific advantages are thought to accrue to US financial institutions. The state, in short, is the political entity, the body politic, that is best able to orchestrate these processes. Failure so to do will likely result in a diminution of the wealth and power of the state.

There is, of course, plenty of uneven geographical development based in part on asymmetrical exchange relations within states. Sub-national political entities, such as metropolitan or regional governments, become critically engaged in such processes. But this is not generally referred to as imperialism. Though some like to talk, with some justification, of internal neocolonialism, or even metropolitan imperialism (on the part of New York or San Francisco), I prefer to leave examination of the role that sub-national regional entities might have in relation to imperialism to a more general theory of uneven geographical development. The effect is to reserve the term imperialism, *pro tem* at least, for a property of inter-state relations and flows of power within a global system of capital accumulation. From the standpoint of capital accumulation, imperialistic politics entails at the very minimum sustaining and exploiting whatever asymmetrical and resource endowment advantages can be assembled by way of state power.

The Logic of Territory and the Logic of Capital

At any given historical–geographical moment, one or other of the logics may dominate. The accumulation of control over territory as an end in itself plainly has economic consequences. These may be positive or negative from the standpoint of exaction of tribute, flows of capital, labour power, commodities, and the like. But this looks quite different to a situation in which territorial control (which may or may not entail actual takeover and administration of territory) is seen as a necessary means to the accumulation of capital. What sets imperialism of the capitalist sort apart from other conceptions of empire is that it is the capitalistic logic that typically dominates, though, as we shall see, there are times in which the territorial logic comes to the fore. But this then poses a crucial question: how can the territorial logics of power, which tend to be awkwardly fixed in space, respond to the open spatial dynamics of endless capital accumulation? And what does endless capital accumulation imply for the territorial logics of power? Conversely, if hegemony within the world system is a property of a state or collection of states, then how can the capitalist logic be so managed as to sustain the hegemon?

Some light is shed on this problem by an acute observation made by Hannah Arendt: 'A never-ending accumulation of property', she wrote, 'must be based on a never-ending accumulation of power. . . . The limitless process of capital accumulation needs the political structure of so "unlimited a Power" that it can protect growing property by constantly growing more powerful.' From this derived, in Arendt's view, 'the "progressive" ideology of the late nineteenth century' which 'foreshadowed the rise of imperialism'.[2] If, however, the accumulation of power must necessarily accompany the accumulation of capital then bourgeois history must be a history of hegemonies expressive of ever larger and continuously more expansive power. And this is exactly what Arrighi records in his comparative history of the shift from the Italian city-states through the Dutch, the British, and now the US phases of global hegemony:

Just as in the late seventeenth and early eighteenth centuries the hegemonic role had become too large for a state of the size of the United Provinces, so in the early twentieth century that role had become too large for a state of the size and resources of the United Kingdom. In both instances, the hegemonic role fell on a state—the United Kingdom in the eighteenth century, the United States in the twentieth century—that had come to enjoy a substantial 'protection rent', that is, exclusive cost advantages associated with absolute or relative geostrategic insularity. . . . But that state in both instances was also the bearer of sufficient weight in the capitalist world economy to be able to shift the balance of power among competing states in

34

whatever direction it saw fit. And since the capitalist world economy had expanded considerably in the nineteenth century, the territory and resources required to become hegemonic in the early twentieth century were much greater than in the eighteenth.[3]

But if Arendt is right, then any hegemon, if it is to maintain its position in relation to endless capital accumulation, must endlessly seek to extend, expand, and intensify its power. But there is, in this, an ever-present danger, for, as Paul Kennedy warns in *The Rise and Fall of the Great Powers*, overextension and overreach have again and again proven the Achilles' heel of hegemonic states and empires (Rome, Venice, Holland, Britain).[4] His warning (in 1990) that the US was itself endangered, if it was heard at all, passed unheeded since, in the decade that has passed since publication of his work, the US has remarkably extended its powers both militarily and politically to a point where the dangers of overreach are palpable. This raises the further question, that if the US is no longer in itself sufficiently large and resourceful to manage the considerably expanded world economy of the twenty-first century, then what kind of accumulation of political power under what kind of political arrangement will be capable of taking its place, given that the world is heavily committed still to capital accumulation without limit? I will return to this question later. But even at this point we can see some intriguing possibilities. Some argue that world government is not only desirable but inevitable. Others argue that some collection of states working in collaboration with each other (in much the way that Kautsky suggested in his theory of ultra-imperialism, and as is hinted at in meetings of organizations such as the G7—now G8) might be able to regulate matters. To this we could add the less optimistic idea that, if it proves impossible for some reason to construct this ever vaster accumulation of political power, then endless capital accumulation will likely dissolve into chaos, ending the era of capital not with a revolutionary bang but in tortured anarchy.

Hegemony

So what constitutes hegemony in the first place? Gramsci's own use of the concept was sufficiently ambiguous to allow multiple interpretations. It sometimes refers solely to political power exercised through leadership and the consent of the governed, as opposed to political power exercised as domination through coercion. On other occasions it seems to refer to the particular mix of coercion and consent embedded in the exercise of political power. I shall have occasion to refer to the latter but interpret hegemony largely in terms of the former. I shall also follow Arrighi's adaptation of the concept to the case of inter-state relations: 'The supremacy of a group or, in this case, a nation state can . . . manifest itself in two ways: as "domination" and as "intellectual and moral leadership". A social group dominates antagonistic groups, which it tends to "liquidate", or to subjugate perhaps even by armed force; it leads kindred or allied groups.' But it can lead in two distinctive ways. By virtue of its achievements, 'a dominant state becomes the "model" for other states to emulate and thereby draws them onto its own path of development. . . . This may enhance the prestige and hence the power of the dominant state . . . but to the extent that emulation is at all successful, it tends to counter-balance and hence deflate rather than inflate the power of the hegemon by bringing into existence competitors and reducing the "specialness" of the hegemon.' Leadership, on the other hand, designates 'the fact that a dominant state leads the system of states in a desired direction and, in so doing, is widely perceived as pursuing a general interest. Leadership in this sense inflates the power of the dominant state.'[5]

An important corollary of this argument is a distinction between 'distributive' and 'collective' power. The former has the character of a zero-sum game in which competition can improve the position of the hegemon by taking power away from others or by leading a regional coalition in some way to bring greater benefits to a region. The recent revival of interest in regional hegemons (the Japanese 'flying geese model', in which Japan leads the rest of Asia, or the European one, in which a Franco-German alliance leads) suggests that this process of redistribution of power is perhaps playing a rather more powerful role in the reorganization of global capitalism than the blanket term 'globalization' tends to imply.[6] But to be truly hegemonic in a global sense entails the use of leadership to create a non-zero-sum game in which all parties benefit, either out of mutual gains from their own interactions (such as trade) or through their enhanced collective power vis-à-vis nature by, for example, the creation and transfer of new technologies, organizational forms, and infrastruc-tural arrangements (such as communication nets and structures of international law). Arrighi emphasizes the accumulation of collective power as the only solid basis for hegemony within the global system. The power of the hegemon, however, is fashioned out of and expressed through an ever-shifting balance between coercion and consensus.

Reflect, for a moment, on how these categories play out in the case of the United States over the last fifty years. The US has frequently relied upon domination and coercion and has not shrunk from the liquidation of opposition. Even internally, it has a history of ruthlessness that belies its attachment to its constitution and the rule of law. McCarthyism, the murder or incarceration of Black Panther leaders, the internment of Japanese in the Second World War, surveillance and infiltration of opposition groups of all kinds, and now a certain preparedness to overthrow the Bill of Rights by passing the Patriot and Homeland Security Acts. It has been even more significantly ruthless abroad in sponsoring coups in Iran, Iraq, Guatemala, Chile, Indonesia, and Vietnam (to name but a few), in which untold thousands died. It has supported state terrorism throughout the world wherever it has been convenient. CIA and special forces units operate in innumerable countries. Study of this record has led many to paint a portrait of the US as the greatest 'rogue state' on earth. There is a major industry in doing so, beginning with Chomsky, Blum, Pilger, Johnson, and many others.[7] While we may only know the half of it, the amazing thing about the US is how much is both known and documented from official or quasi-official sources and what a grizzly, despicable, and deeply disturbing record it is. Liquidation can come by a variety of means. The economic power to dominate (such as the trade embargo on Iraq and Cuba or IMF austerity programmes implemented at the behest of the US Treasury) can be used with equally destructive effect as physical force. The distinctive role of US financial institutions and the US Treasury backed by the IMF in visiting a violent devaluation of assets throughout East and South-East Asia, creating mass unemployment and effectively rolling back years of social and economic progress on the part of huge populations in that region, is a case in point. Yet most of the US population either lives in a state of denial, refusing even to hear of such things, or, if it does hear, passively accepts liquidations and coercions as facts of life, the normal cost of doing fundamentally honest business in a dirty world.

But what the critics who dwell solely on this aspect of US behaviour in the world all too often fail to acknowledge is that coercion and liquidation of the enemy is only a partial, and sometimes counterproductive, basis for US power. Consent and cooperation are just as important. If these could not be mobilized internationally and if leadership could not be exercised in such a way as to generate collective benefits, then the US would long ago have ceased to be hegemonic. The US must at least act in such a way as to make the claim that it is acting in the general interest plausible to others even when, as most people suspect, it is acting out of narrow self-interest. This is what exercising leadership through consent is all about.

In this regard, of course, the Cold War provided the US with a glorious opportunity. The United States, itself dedicated to the endless accumulation of capital, was prepared to accumulate the political and military power to defend and promote that process across the globe against the communist threat. Private property owners of the world could unite, support, and shelter behind that power, faced with the prospect of international socialism. Private property rights were held as a universal value and proclaimed as such in the UN Declaration of Human Rights. The US guaranteed the security of European democracies, and benevolently helped rebuild the war-torn economies of Japan and West Germany. Through its policy of 'containment' it tacitly established the boundaries of its own informal empire (particularly in Asia), while committing itself to undermining by whatever means possible the power of its great competitor, the Soviet empire. While we know enough about decision-making in the foreign policy establishment of the Roosevelt–Truman years and since to conclude that the US always put its own interests first, sufficient benefits flowed to the propertied classes in enough countries to make US claims to be acting in the universal (read 'propertied') interest credible and to keep subaltern groups (and client states) gratefully in line. This 'benevolence' is quite plausibly presented by defenders of the US in response to those who emphasize the rogue state image based in coercion. It is also heavily emphasized in the way in which the US typically views and presents itself to the rest of the world, though here there is as much myth-spinning as truth-telling. The US likes to believe, for example, that it and it alone liberated Europe from the Nazi yoke, and it erases entirely the much more important role of the Red Army and of the siege of Stalingrad in turning the tables in the Second World War. The more general truth is that the US engages in both coercive and hegemonic practices simultaneously, though the balance between these two facets in the exercise of power may shift from one period to another and from one administration to another.

How America's Power Grew

The US has for many years definitely exercised leadership of that part of the world dedicated to endless capital accumulation and consequently spread its ways of doing business far and wide. It did not, of course, during the Cold War years, exercise a truly global hegemony. With the threat of communism now effectively gone, the US leadership role is harder to define and sustain. This is the question that is being not so subtly debated by those who wish to project the future of US imperialism and empire upon the world in the twenty-first century. This is also the question being asked by those who see a regional partition of powers as an alternative configuration of political arrangements within the overall rules of neo–liberal globalization.

There is no question either, that emulation has played an important role in global affairs. Much of the rest of the world has been entrained politically, economically, and culturally in globalization through Americanization. But here I depart somewhat from Arrighi, since I cannot see that emulation always creates competition and that it is always a zero-sum game. The emulation of US consumerism, ways of life, cultural forms, and political and financial institutions has contributed to the process of endless capital accumulation globally. Situations may indeed arise where emulation leads to sharpened competition (as, for example, when Taiwan totally takes over some sector of production from the US). And this can have major impacts upon the domestic situation in the United States (as the long history of deindustrialization in arenas such as steel, shipbuilding, and textiles within the United States illustrates). But I think it important to distinguish between this and other aspects of emulation that

41

actually contribute to the formation of greater collective powers.

Political power is always constituted out of some unstable mix of coercions, emulations, and the exercise of leadership through the development of consent. These are the means. But what of the forms of power that must be amassed within the territorial logic to ensure its ability to realize its interests? The intangibles of prestige, status, deference, authority, and diplomatic clout must be grounded materially in something. Money, productive capacity, and military might are the three legs upon which hegemony stands under capitalism. But here, too, we find shifting and unstable configurations. Consider, as an example, the shifting material bases of US hegemony since the end of the nineteenth century.

The Rise of Bourgeois Imperialisms, 1870–1945

Arendt asserts that the imperialism that arose towards the end of the nineteenth century was 'the first stage in the political rule of the bourgeoisie rather than the last stage of capitalism'.[8] The evidence for this is substantial. The first major crisis of capitalist overaccumulation (defined primarily as a surplus of capital lacking profitable means of employment—but see Chapter 3 for a more extensive treatment) was the Europe-wide economic collapse of 1846–50 that sparked bourgeois revolutionary movements (with more than a hint of working-class participation) all over Europe. The partial incorporation of the bourgeoisie within the state apparatus thereafter proceeded unevenly across Europe. The way out of this first capitalist crisis was a double movement of long-term infrastructural investments (of the sort laid out in the theory of 'productive state expenditures' that underlay Haussmann's transformation of Paris and the widespread attention given to transportation, water, and sewage programmes and investment in housing and public facilities in many other European countries) and geographical expansions particularly focused on Atlantic trade (with the US a primary outlet). But by the mid-1860s the ability to absorb capital and labour surpluses by these means was running out. The interruption of the Atlantic trade by the American Civil War had a serious impact, and internal political movements (of the sort that produced the Paris Commune of 1871) were creating internal stresses across Europe. In the aftermath of the Civil War, proletarian movements arose in the United States as well.

Surplus capitals in Europe, increasingly blocked by assertive capitalist class power from finding internal uses, were forced outwards to swamp the world in a massive wave of speculative investment and trade, particularly after 1870 or so. The capitalistic logic of searching for what, in Chapter 3, I will call 'spatio-temporal fixes' surged to the forefront on a global scale. The need to protect these foreign ventures and even to regulate their excesses put pressure on states to respond to this expansionary capitalistic logic. For that to occur required that the bourgeoisie, which already held power in the United States, consolidate its political power vis-à-vis older class formations and either dissolve older imperialist forms (such as that of the Austro-Hungarian or Ottoman empires) or convert them (as in Britain) to a distinctively capitalistic logic. The consolidation of bourgeois political power within the European states was, therefore, a necessary precondition for a reorientation of territorial politics towards the requirements of the capitalistic logic.

The bourgeoisie had, however, appealed to the idea of nation in its ascent to power. The wave of nation-state formation that occurred during the latter half of the nineteenth century in Europe (in Germany and Italy in particular) logically pointed to a politics of internal consolidation rather than to foreign ventures. Furthermore, the political solidarity supposed by the idea of nation could not easily be extended to those who are 'others' without diluting what the idea of nation is supposed to represent. The nation-state does not in itself, therefore, provide a coherent basis for imperialism. How, then, could the problem of overaccumulation and the necessity of a global spatio-temporal fix find an adequate political response on the basis of the nation-state? The answer was to mobilize nationalism, jingoism, patriotism, and, above all, racism behind an imperial project in which national capitals—and at this time there was a plausible coherence between the scale of capitalist enterprise and the scale on which nation-states were working—could take the lead. This, as Arendt points out, meant the suspension of internal class struggle and the construction of an alliance between what she calls 'the mob' and capital within the nation-state. 'So unnatural did this seem in Marxist terms,' she observes, ' that the actual dangers of the imperialist attempt—to divide mankind into master and slave races, into higher and lower breeds, into colored and white men, all of which were attempts to unify the people on the basis of the mob—were completely overlooked.' There may be, she says, 'an abyss between nationalism and imperialism' in theory, 'but in practice, it can and has been bridged by tribal nationalism and outright racism'.[9] That this would actually be the outcome was not of course inevitable. But the struggle against it ultimately failed, as was shown most dramatically with the Second Socialist International's collapse as each national branch fell in line in support of its country in the 1914–18 war. The consequences were quite horrifying. A variety of nation-based and therefore racist bourgeois imperialisms evolved (British, French, Dutch, German, Italian). Industrially driven but non-bourgeois imperialisms also arose in Japan and Russia. They all espoused their own particular doctrines of racial superiority, given pseudo-scientific credibility by social Darwinism, and more often than not came to view themselves as organic entities locked in a struggle for survival with other nation-states. Racism, which had long lurked in the wings, now moved to the forefront of political thinking. This conveniently legitimized the turn to what in Chapter 4 I will call 'accumulation by dispossession' (of barbarians, savages, and inferior peoples who had failed to mix their labour properly with the land) and the extraction of tribute from the colonies in some of the most oppressive and violently exploitative forms of imperialism ever invented (the Belgian and Japanese forms being perhaps the most vicious of all). It is, as Arendt argues, also important to see Nazism and the Holocaust as something that is entirely comprehensible though by no means determined within this historical-geographical trajectory.

The underlying contradiction between bourgeois nationalism and imperialism could not be resolved, while the rising need to find geographical outlets for surplus

45

capitals put all manner of pressures on political power within each imperialist state to expand geographical control. The overall result, as Lenin so accurately predicted, was fifty years of inter-imperialist rivalry and war in which rival nationalisms featured large. Its essential features involved the carving up of the globe into distinctive terrains of colonial possession or exclusionary influence (most dramatically in the grab for Africa of 1885 and the Versailles settlement after the First World War, including its partitioning of the Middle East between French and British protectorates); the pillaging of much of the world's resources by the imperial powers; and the widespread deployment of virulent doctrines of racial superiority; all matched by a total and predictable failure to deal with the surplus capital problem within closed imperial domains, as seen in the great depression of the 1930s. Then came the ultimate global conflagration of 1939–45.

Although the early phases were marked by British hegemony and at least a modicum of free trade, I think Arendt is right to see the period from 1870 to 1945 as cut from exactly the same cloth of rival nation-based imperialisms that could only work through the mobilization of racism and the construction of national solidarities favourable to fascism at home and prone to violent confrontation abroad.

In the midst of all of this, the US was evolving its own distinctive form of imperialism. Powered by a remarkable spurt of capitalist development after the Civil War, the US was becoming technologically and economically dominant vis-à-vis the rest of the world. Its governmental form, not burdened with feudal or aristocratic residuals of the sort to be found in Europe, broadly reflected corporate and industrial class interests and had, ever since independence, been bourgeois to the core (as formalized in its Constitution). Political power internally was devoted to individualism and bitterly opposed to any threat to the inalienable rights of private property and the profit rate. It was a multi-ethnic immigrant society which made narrow ethnic nationalism of the sort found in Europe and Japan impossible. It was also exceptional in possessing abundant space for internal expansion, within which both the capitalistic and political logics of power could find room for manoeuvre. Its own internalized form of racism (towards blacks and indigenous peoples) was paralleled by an antagonism to 'non-Caucasians' more generally that curbed the temptation to absorb territories (such as that of Mexico or in the Caribbean) where non-Caucasian populations dominated. The theory of manifest destiny fuelled its own particular brand of expansionary racism and international idealism. From the late nineteenth century onwards, the US gradually learned to mask the explicitness of territorial gains and occupations under the mask of a spaceless universalization of its own values, buried within a rhetoric that was ultimately to culminate, as Neil Smith points out, in what came to be known as 'globalization'.[10] The United States had phases of emulating the Europeans, had episodic moments when it seemed that geographical expansion was economically essential and it had long declared, through the various formulations of the Monroe Doctrine, that the whole of the Americas should be free of European control and therefore de facto within its own sphere of domination. And it was Woodrow Wilson's dream to make the Monroe Doctrine universal. But in South America the US encountered republics that, like itself, had freed themselves from the colonial yoke through independence struggles. It therefore had to work out means of imperial domination that nominally respected the independence of such countries yet dominated them through some mix of privileged trade relations, patronage, clientelism, and covert coercion. While the US generally held to the principle of the 'open door' with respect to global trade it had, however, little inclination or real means to enforce it before the Second World War. It became involved in the First World War, played an important role in shaping the Versailles settlement, in which the principle of national self-determination was at least recognizable, though not practised (particularly with respect to the Middle East), experienced the trauma of the Great Depression (more a result of internal failures of class rule than a reflection of lack of opportunities for US-based capital to expand geographically), and was drawn into the subsequent global conflicts spawned by inter-imperialist rivalries. But with strong isolationist currents on both the left and right and a long historical fear of foreign entanglements as inimical to its own form of governance, imperial thrusts were occasional and limited, mainly covert rather than overt, politically rather than capitalistically motivated, except in the case of individual corporations with particular foreign interests that shamelessly mobilized political power to back their specific projects whenever and wherever necessary. The US was still as much a potential absorber as a producer of surplus capital, though in the 1930s it failed entirely to realize its own potentialities in this regard, in large part because of the internal configuration of class power that resisted even Roosevelt's modest attempts during the New Deal to rescue the economy from its contradictions through redistributions of wealth. The difficulty of achieving internal cohesion in an ethnically mixed society characterized by intense individualism and class division also produced what Hofstadter calls 'the paranoid style' of American politics: fear of some 'other' (such as bolshevism, socialism, anarchism, or merely 'outside agitators') became crucial to creating political solidarities on the home front.[11] The Soviet Union and bolshevism were increasingly cast in the role of chief enemies and villains (with fear of China, including Chinese immigration, lurking in the wings).

The Post-War History of American Hegemony, 1945–1970

The US emerged from the Second World War as by far the most dominant power. It dominated in technology and production. The dollar (backed by most of the world's gold supply) was supreme, and its military apparatus was far superior to any other. Its only serious opponent was the Soviet Union, but that country had lost vast numbers of its population and suffered terrible degradation of its military and industrial capacity compared to the United States. It had borne the brunt of the fighting against Nazism and, arguably, the siege of Leningrad and the subsequent destruction of much of Germany's military capacity on the eastern front was crucial to the Allied victory. The delay in launching a second front in Europe infuriated Stalin and may in itself have been calculated by the US and Britain as a means to let the Soviet Union bear the brunt of the fighting. But the delay had serious consequences since it permitted the Soviet Union to make major territorial gains in Europe from which it subsequently refused to retreat, installing client regimes throughout eastern Europe, even into East Germany. For the Soviet Union defence of its interests amounted to defence of its territorial control.

During the war, elite elements within the US government and the private sector outlined a post-war settlement plan that would guarantee peace, economic growth, and stability. Territorial aggrandizement was ruled out. It had long been an influential principle of political thought and practice in the United States, from James Madison onwards, that foreign entanglements should be avoided because they would undermine democracy at home. The difficulty was to bridge the gap between this fear and the obvious fact of US global domination. Much as European imperialism had turned to racism to bridge the tension between nationalism and imperialism, so the US sought to conceal imperial ambition in an abstract universalism. The effect, as Neil Smith observes, was to deny the significance of territory and geography altogether in the articulation of imperial power. This was the move that Henry Luce made in his influential 1941 cover editorial in *Life* magazine entitled 'The American Century'. Luce, an isolationist, considered that history had conferred global leadership on the United States and that this role, though thrust upon it by history, had to be actively embraced. The power conferred was global and universal rather than territorially specific, so Luce preferred to talk of an American century rather than an empire. Smith remarks:

Whereas the geographical language of empires suggests a malleable politics—empires rise and fall and are open to challenge—the 'American Century' suggests an inevitable destiny. In Luce's language, any political quibble about American dominance was precluded. How does one challenge a century? US global dominance was presented as the natural result of historical progress, implicitly the pinnacle of European civilization, rather than the competitive outcome of political-economic power. It followed as surely as one century after another. Insofar as it was beyond geography, the American Century was beyond empire and beyond reproof.[12]

The fact of Soviet territorial gains and burgeoning power ran up against 'the paranoid style' of US politics to produce the Cold War. Internally this led to the repressions known as 'McCarthyism' which curbed freedoms of expression and fiercely opposed anything that sounded remotely communistic or socialistic. The unions were purged of radical influences, and communist and other leftist parties were effectively proscribed. The FBI infiltration of anything considered oppositional began in earnest. All of this was legitimized as vital to the internal security of the United States in the face of the Soviet threat. The result was political conformity and solidarity at home. Leviathan, as Arendt might put it, imposed order upon the potential chaos of individual interests. Labour was pushed and cajoled into a general compact with capital, coupling wages with productivity gains (a Fordist model considered worthy of emulation). Working-class support was procured for US politics abroad in the name of anti-communism and economic self-interest.

In foreign affairs, the US presented itself as chief defender of freedom (understood in terms of free markets) and of the rights of private property. The US provided economic and military protection for propertied classes or political/military elites wherever they happened to be. In return these propertied classes and elites typically centred a pro-American politics in whatever country they happened to be. This implied military, political, and economic containment of the sphere of influence of the Soviet Union.[13] The imperial realm of the United States was defined negatively, as everything not directly contained in the Soviet orbit (which in US eyes included China long after it had gone its separate way). While it was accepted that frontal confrontation with the Soviet empire was impossible, every opportunity was seized to undermine it—a policy that led into some disasters as the US supported the rise of the Mujahidin and Islamic fundamentalism in order to embarrass the Soviets in Afghanistan, only to have to suppress the Mujahidin's influence later in a war against terrorism based in Islamic fundamentalism. Any expansion of communist-controlled territory was viewed as a serious loss—hence the intense recriminations over 'who lost China' to Mao and the use of that accusation to spearhead McCarthy's attacks.

Two cardinal principles of internal strategic practice had been defined during the Second World War, and these remained set in stone thereafter: the social order in the United States should remain stable (no radical redistributions of wealth or power and no challenge to elite and/or capitalist class control would be tolerated), and there should be a continuous expansion of domestic capital accumulation and consumption to ensure domestic peace, prosperity, and tranquillity.[14] Foreign engagements should not interfere with consumerism at home: hence the preference for what Ignatieff calls 'empire lite'. The United States would use its superior military power to protect client regimes throughout the world that were supportive of US interests. The overthrow of Mossadegh, who had nationalized the oil fields of Iran, and his replacement by the Shah in 1953 (all with CIA help) and the subsequent reliance upon him to look out for US interests in the Gulf region was typical of this approach. In key geopolitical arenas, such as the frontline states with the Soviet Union, it would use its economic might to build strong economies based on capitalistic principles (hence the Marshall Plan for Europe and strong support for Japan, Taiwan, South Korea, and other vulnerable frontline states in relation to Soviet power). Access to the Middle East, with its oil reserves, was also crucial (Roosevelt, though sick, went out of his way to stop off to talk with the Saudis and others about the importance of maintaining the flows of oil on his way back from the Yalta conference).

The US placed itself at the head of collective security arrangements, using the United Nations and, even more importantly, military alliances such as NATO, to limit the possibility of inter-capitalist wars and to combat the influence of the Soviet Union and then China. It used its own military power, covert operations, and all manner of economic pressures to ensure the creation or continuance of friendly governments. To this end it was prepared to support the overthrow of democratically elected governments and to engage directly or indirectly in tactics of liquidation of those considered opposed to US interests. It did so in Iran, Guatemala, Brazil, the Congo, the Dominican Republic, Indonesia, Chile, and elsewhere. It intervened electorally and covertly in dozens of other countries throughout the world. Yet it lost out in China and Cuba, and communist insurgencies thrived elsewhere as the Soviet model gained traction as a means to bring about rapid modernization without capitalist class rule.

Within the 'free world' the US sought to construct an open international order for trade and economic development and rapid capital accumulation along capitalistic lines. This required the dismantling of the former nation-state-based empires. Decolonization required state formation and self-governance across the globe. The US largely modelled its relationships with these newly independent states on its experience in dealing with the independent republics of Latin America during the pre-war period. Privileged trade relations, clientelism, patronage, and covert coercion were, as we have seen, the chief weapons of control. And the US deployed these weapons bilaterally, country by country, thus positioning itself as a central hub with innumerable spokes connecting it to all other states around the world. Any threat of collective action against overwhelming US power could be countered by a divide-and-rule strategy making use of individual connections to limit collective autonomy, even when, as in Europe, moves towards union were under way.

An international framework for trade and economic development within and between these independent states was set up through the Bretton Woods agreement to stabilize the world's financial system, accompanied by a whole battery of institutions such as the World Bank, the International Monetary Fund, the International Bank of Settlements in Basle, and the formation of organizations such as GATT (the General Agreement on Tariffs and

Trade) and the OECD (Organization for Economic Cooperation and Development), designed to coordinate economic growth between the advanced capitalist powers and to bring capitalist-style economic development to the rest of the non-communist world. In this sphere the US was not only dominant but also hegemonic in the sense that its position as a super-imperialist state was based on leadership for propertied classes and dominant elites wherever they existed. Indeed, it actively encouraged the formation and empowerment of such elites and classes throughout the world: it became the main protagonist in projecting bourgeois power across the globe. Armed with Rostow's theory of 'stages' of economic growth, it strove to promote the 'take-off' into economic development that would promote the drive to mass consumption on a country-by-country basis in order to ward off the communist menace.[15]

But the dismantling of European-based imperialisms also entailed the formal disavowal of the racism that had permitted the reconciliation of nationalism with imperialism. The UN Declaration of Human Rights and various UNESCO studies denied the validity of racism and sought to found a universalism of private property and of individual rights that would be appropriate for a second stage of bourgeois political rule. For this to work demanded that the US should depict itself as the pinnacle of civilization and a bastion of individual rights. Pro-Americanism had to be cultivated and projected abroad. And so began the huge cultural assault upon 'decadent' European values and the promotion of the superiority of American culture and of 'American values'. Money power was used to dominate cultural production and influence cultural values (this was the era when New York 'stole' the idea of modern art from Paris[16]). Cultural imperialism became an important weapon in the struggle to assert overall hegemony. Hollywood, popular music, cultural forms, and even whole political movements, such as those of civil rights, were mobilized to foster the desire to emulate the American way. The US was constructed as a beacon of freedom that had the exclusive power to entrain the rest of the world into an enduring civilization characterized by peace and prosperity.

But the US also came to be viewed as the primary engine of capital accumulation and one that could entrain the rest of the world in its tracks. Massive internal transformations in its own economy (that had been merely hinted at during the New Deal of the 1930s) became of great global importance because of the market opportunities it spawned. Investments in education, the interstate highway system, sprawling suburbanization, and the development of the south and west, absorbed vast quantities of capital and product in the 1950s and 1960s. The US state, to the chagrin of neo-liberals and conservatives, became a developmental state during these years. Except for a few key areas, such as strategic resources, the US did not rely too much on the extraction of value from the rest of the world. The proportion of GDP growth attributable to foreign trade remained less than 10 per cent up until the 1970s. While there were some foreign operations, like ITT (International Telephone and Telegraph) in Chile (one of whose directors had been director of the CIA), or United Fruit in Central America, which exercised considerable influence over US foreign policy in those regions, US economic imperialism was, with the exception of strategic minerals and oil, rather muted. In so far as an outer dialectic was called for, it pointed to the already developed parts of the capitalist world. Direct foreign investment flowed to Europe, leading Europeans to become obsessed with holding off what Servan-Schreiber called 'the American challenge'.[17] In return, however, the US opened its market to others and provided an effective demand for products from Europe and Japan. Strong growth occurred throughout the capitalist world. The accumulation of capital proceeded apace through 'expanded reproduction'. Profits were reinvested in growth as well as in new technologies, fixed capital, and extensive infrastructural improvements.[18] Controls over capital outflows (as opposed to commodities) were, however, retained from the preceding period, particularly in Europe. This gave individual states considerable discretion over fiscal as well as monetary policies. The role of financial speculation remained relatively muted and territorially confined. This 'Keynesian' context for state expenditures cohered with a dynamic of class struggle within individual nation-states over distributive questions. This was an era when organized labour became quite strong and social democratic welfare states emerged across Europe. The social wage became an object of struggle even within the United States, and organized labour won several significant victories internally over wage levels and living standards.

The period from 1945 to 1970 was, then, the second stage in the political rule of the bourgeoisie operating under global US dominance and hegemony. It brought a period of remarkably strong economic growth to the advanced capitalist countries. A tacit global compact was established among all the major capitalist powers, with the US in a clear leadership role, to avoid internecine wars and to share in the benefits of an intensification of an integrated capitalism in the core regions. The geographical expansion of capital accumulation was assured through decolonization and 'developmentalism' as a generalized goal for the rest of the world. Expanded reproduction seemed to be working very well and secondary effects even spilled outwards, though lightly and unevenly, across the non-communist world. Internally, the increasing power of labour within the capital–labour pact meant spreading the benefits of consumerism to the lower classes, even to some minorities (though not enough, as the urban unrest of the 1960s proved). The problem of overaccumulation of capital, though always threatening, was contained until the late 1960s by a mix of internal adjustments and spatio-temporal fixes both within and without the United States. These strategies, it was hoped, would permit the system to overcome the economic problems that had plagued the 1930s and protect against the threat of communism.

But this second stage was not free of contradictions. First, the formal disavowal of racism internationally posed all manner of difficulties internally for the United States, where racial discrimination was rampant. The civil rights movement which, in the end, provided a model for much of the rest of the world, had its origins in internal dynamics, as did the urban uprisings led by blacks in the 1960s; but it also had an international dimension as the universalism of human rights clashed with internal practices and as 'coloured' diplomats en route between the UN in New York and Washington, DC found themselves barred from staying in motels. The racial selectivity of US immigration policy also came under fire. Migrant flows into the US began to change their character.

Secondly, as we will see in Chapter 3, the policy of an open market made the US vulnerable to international competition. Capital flows during this period were heavily concentrated within the advanced capitalist world (broadly within the OECD countries). West Germany and Japan in particular ratcheted up their economic power to challenge US dominance in production during the 1960s. As the ability of the US to absorb surplus capitals internally began to flag in the late 1960s, so overaccumulation emerged as a problem and economic competition sharpened.

Thirdly, whenever there was a conflict between democracy, on the one hand, and order and stability built upon propertied interests, on the other, the US always opted for the latter. The US therefore moved from the position of patron of national liberation movements to oppressor of any populist or democratic movement that sought even a mildly non-capitalist (let alone a socialist or communist) path to the improvement of economic well-being. Social democratic or populist attempts at modifying capitalism were often ruthlessly struck down (as happened to Bosch in the Dominican Republic, Goulart in Brazil, and, eventually, Allende in Chile). Even in Europe the US did everything in its power to undermine socialism and even on occasion to subvert social democracy. And savagely dictatorial regimes, such as those in Argentina in the 1970s, the Saudis, the Shah of Iran, and Suharto in Indonesia, were unconditionally supported by US military and economic power since they supported US interests. Growing resentment of being locked into a spatio-temporal situation of perpetual subservience to the centre also sparked anti-dependency movements throughout the developing world. Class and national liberation struggles within the developing world were more and more forced into an anti-American politics. Anti-dependency fused with anti-colonialism to define anti-imperialism. In all of these struggles the territoriality of political power was just as important to the sustenance of US hegemony as it had been to the European empires that went before. The US did not acquire its imperial stature, as Ignatieff avers, through denial: it simply used denial of geography and the rhetoric of universality to hide its territorial engagements, more so from itself than from others.

Fourthly, the effect of the Cold War and of these foreign entanglements was to empower what President Eisenhower critically referred to in his farewell address as a dangerously powerful 'military industrial complex'. This threatened to dominate politics through its pervasive influence and pursue its own narrow interests by exaggerating threats and manipulating external crises so as to construct a permanent war economy that would render it ever more powerful. To survive economically, the defence industries needed a thriving export trade in arms. This came to have a fundamental role in US capital accumulation, but it also resulted in the excessive militarization of the rest of the world.

This second stage in global rule of the bourgeoisie came to an end around 1970 or so. The problems were multiple. First there was the classic problem of all imperial regimes—overreach. The containment of (and attempt to subvert) communism proved rather more costly than expected for the United States. The rising costs of the military conflict in Vietnam, when coupled with the golden rule of never-ending domestic consumerism—a policy of guns and butter—proved impossible to sustain, since military expenditures provide only short-run outlets for surplus capital and generate little in the way of long-term relief to the internal contradictions of capital accumulation. The result was a fiscal crisis of the developmental state within the United States. The immediate response was to use the right of seigniorage and print more dollars.[19] This resulted in world-wide inflationary pressures. The consequence, as we shall see in Chapter 3, was an explosion in the quantity of 'fictitious' capital in circulation lacking any prospect of redemption, a wave of bankruptcies (focused initially on assets in the built environment), uncontainable inflationary pressures, and the collapse of the fixed international arrangements that had founded US super-imperialism after the Second World War. Meanwhile, the growing power of organized labour throughout the core states of the global system pushed up the level of social expenditures as well as wage costs, thus cutting into profits. Stagflation resulted. Profit opportunities disappeared and a crisis of overaccumulation of capital emerged. The debt overhang of many governments from vast investments in physical and social infrastructures produced a fiscal crisis of the state (culminating in the spectacular bankruptcy of New York City in 1975). To top it all, the competitive strength of the revived Japanese and West German industries challenged, and in some areas now surpassed, US dominance in production. Emulation in manufacturing was cutting off one of the key legs of US hegemony. The United States' economic position seemed untenable. Surplus dollars flooded the world market and the whole financial architecture of the Bretton Woods system collapsed.

Neo-liberal Hegemony, 1970–2000

A different kind of system then emerged, largely under US tutelage. Gold was abandoned as the material basis of money values and thereafter the world had to live with a dematerialized monetary system. Flows of money capital, already moving freely around the world via the eurodollar market (dollars held outside the United States that could easily be lent anywhere) were to be totally liberated from state controls. The collusion (now documented) between the Nixon administration and the Saudis and Iranians to push oil prices sky-high in 1973 did far more damage to the European and Japanese economies than it did to the US (which at that time was not greatly dependent upon Middle Eastern supplies). US banks (rather than the IMF, which was the preferred agent of the other capitalist powers) gained the monopoly privilege of recycling the petrodollars into the world economy, thus bringing the eurodollar market back home.[20] New York became the financial centre of the global economy (this, coupled with internal deregulation of financial markets, allowed that city to recover from its crisis and to flourish to the point of incredible affluence and conspicuous consumption in the 1990s).

Threatened in the realm of production, the US had countered by asserting its hegemony through finance. But for this system to work effectively, markets in general and capital markets in particular had to be forced open to international trade (a slow process that required fierce US pressure backed by use of international levers such as the IMF and an equally fierce commitment to neo-liberalism as the new economic orthodoxy). It also entailed shifting the balance of power and interests within the bourgeoisie from production activities to institutions of finance capital. Financial power could be used to discipline working-class movements. The opportunity arose to launch a frontal assault on the power of labour and to diminish the role of its institutions in the political process. President Reagan's first move was to destroy the strong collective power of the air traffic controllers (PATCO), and this served notice on the union movement that it stood to suffer the same fate should any other group of workers strike. A wave of labour militancy swept the advanced capitalist world during the late 1970s and the 1980s (the miners taking the lead in both Britain and the United States) as working-class movements everywhere sought to preserve the gains they had won during the 1960s and early 1970s. In retrospect, we can see this as a rearguard action to preserve conditions and privileges gained within and around expanded reproduction and the welfare state, rather than a progressive movement seeking transformative changes. For the most part this rearguard action failed. The subsequent devaluation of labour power and the steady relative degradation in the condition of the working class in the advanced capitalist countries was then paralleled by the formation of a huge, amorphous, and unorganized proletariat throughout much of the developing world. This put downward pressure upon wage rates and labour conditions everywhere. Easily exploited low-wage workforces coupled with increasing ease of geographical mobility of production opened up new opportunities for the profitable employment of surplus capital. But in short order this exacerbated the problem of surplus capital pro-duction world-wide. Nevertheless, unemployment surged and wage rates and working-class militancy were held in check. The debt overhang of the state opened up all man-ner of opportunities for speculative activity that, in turn, made state powers more vulnerable to financial influences. Finance capital, in short, moved centre-stage in this phase of US hegemony, and it was able to exercise a certain dis-ciplinary power over both working-class movements and state actions, particularly whenever and wherever the state ran up significant debts.

This whole shift would not have had the effect it did had it not been for a battery of technological and organ-izational shifts that allowed manufacturing to become much more footloose and flexible. Reductions in the cost of transport, coupled with political shifts on the part of governments at all levels to offer a positive business climate and to cover some of the fixed costs of relocation, promoted the kind of geographical mobility of manufac-turing capital that the increasingly hyper-mobile financial capital could feed upon. While the shift towards financial power brought great direct benefits to the United States, the effects upon its own industrial structure were nothing short of traumatic, if not catastrophic. Offshore produc-tion became possible and the search for profit made it probable. Wave after wave of deindustrialization hit industry after industry and region after region within the US, beginning with the low-value-added goods (such as textiles), but step by step ratcheting up the value-added scale through sectors such as steel and shipbuilding to high-tech imports, particularly from East and South-East Asia. Even Chrysler had to be bailed out (effectively nationalized for a short period) by the Federal Government to avoid closure. The US was complicit in undermining its dominance in manufacturing by unleashing the powers of finance throughout the globe. The benefit, however, was ever cheaper goods from elsewhere to fuel the endless consumerism to which the US was committed. US dependency on foreign trade was on the rise and the need to build and protect asymmetrical trade relations moved to the fore as a key objective of political power.

By 1980 or so it became clear that manufacturing in the United States was now but one complex among many operating in a highly competitive global environment, and that the only way it could survive was by achieving superiority (usually temporary) in productivity and in product design and development. It was, in short, no longer hegemonic. It needed help from government (such as the Plaza accord of 1985 in which government agreed to depreciate the dollar against the yen to make US manufacturing exports more competitive—a tactic that had to be reversed in the 1990s as Japanese manufacturing stagnated). Some special sectors—agribusiness and defence for example—were immune, but the rest were forced into radical adjustments in everything from techniques of production to labour relations. In those areas where US firms remained powerful, the turn to offshore production of components or even whole products placed more and more productive capacity outside the borders of the United States even though the repatriation of profits kept wealth flowing towards it. In other areas, the monopoly privileges that attach to patented technologies and licensing laws gave welcome relief from the draining away of US dominance in production. The US was moving towards becoming a rentier economy in relation to the rest of the world and a service economy at home. But sufficient wealth accrued to continue the consumerism that had always been the basis of social peace.

Internationally, finance capital proved more and more volatile and predatory. Various bouts of devaluation and destruction of capital were visited (usually through the good graces of IMF structural adjustment programmes) as an antidote to the inability to keep capital accumulation going smoothly by expanded reproduction. In some instances, for example in Latin America in the 1980s, whole economies were raided and their assets recovered by US finance capital. In others, it was more simply an export of devaluation. The hedge funds' attack upon the Thai and Indonesian currencies in 1997, backed up by the savage deflationary policies demanded by the IMF, drove even viable concerns into bankruptcy throughout East and South-East Asia. Unemployment and impoverishment were the result for millions of people. That crisis also conveniently sparked a flight to the dollar, confirming Wall Street's dominance and generating an amazing boom in asset values for the affluent in the United States. Class struggles began to coalesce around issues such as IMF-imposed structural adjustment, the predatory activities of finance capital, and the loss of rights through privatization. The tone of anti–imperialism began to shift towards antagonism to the main agents of financialization—the IMF and the World Bank being frequently singled out.

Debt crises within particular countries (two-thirds of IMF members experienced a financial crisis after 1980, some more than twice) could be used, however, to reorganize the internal social relations of production in each country where they occurred in such a way as to favour the further penetration of external capitals.[21] Domestic financial regimes, domestic product markets, and thriving domestic firms were, in this way, prised open for takeover by American, Japanese, or European companies. Low profits in the core regions could thereby be supplemented by taking a cut out of the higher profits being earned abroad. What I call 'accumulation by dispossession' (see Chapter 4), became a much more central feature within global capitalism (with privatization as one of its key elements). Resistance in this sphere, rather than through the labour struggles typically spawned by expanded reproduction, became more central within the anti-capitalist and anti-imperialist movement.

While centred on the Wall Street–Treasury complex, the system had many multilateral aspects. The financial centres of Tokyo, London, Frankfurt, and many other places took part in the action as financialization cast its net across the world, focusing on a hierarchically ordered set of financial centres and a transnational elite of bankers, stockbrokers, and financiers. This was associated with the emergence of transnational capitalist corporations which, though they may have had a basis in one or other nation-state, spread themselves across the map of the world in ways that were unthinkable in earlier phases of imperialism (the trusts and cartels that Lenin and Hilferding described were all tied very closely to particular nation-states). This was the world that the Clinton White House, with an all-powerful Treasury Secretary, Robert Rubin, drawn from the speculator side of Wall Street, sought to manage by a centralized multilateralism (epitomized by the so-called 'Washington Consensus' of the mid-1990s). The multilateralism was increasingly organized around a regionalization of the global economy with a triadic structure of North America (NAFTA), Europe (the EU), and the looser confederation of interests built around trading relations in East and South-East Asia dominating. With the neo-liberal ground rules of open financial markets and relatively free access being strengthened, there seemed little danger of these regional configurations lapsing back into the competitive autarky that had proven so destructive in the period before the Second World War and which had played such an important role in laying the basis for inter-capitalist war. Within this triadic structure, however, it seemed clear that the US still held the major cards by virtue of its huge consumer market, its overwhelming financial power, and its reserve of unchallenged military might.

And, to top it all, the end of the Cold War suddenly removed a long-standing threat to the terrain of global capital accumulation. The collective bourgeoisie had indeed inherited the earth. Fukuyama prophesied that the end of history was at hand. It seemed, for a brief moment, that Lenin was wrong and that Kautsky might be right— an ultra-imperialism based on a 'peaceful' collaboration between all the major capitalist powers (now symbolized by the grouping known as the G7, expanded to the G8 to incorporate Russia, albeit under the hegemony of US leadership) was possible—and that the cosmopolitan character of finance capital (symbolized by the meetings of the World Economic Forum in Davos) would be its founding ideology.[22]

But it would be wrong to think of this financial power, awesome though it definitely was, as being omnipotent and able to impose its will without constraint. It is in the very nature of financialization to be perpetually vulnerable in relationship to the production of value in industrial and agricultural activity. In the midst of all the raiding and devaluation, there arose new and significant complexes of industrial production. In East and South-East Asia, for example, regional complexes such as the Pearl River delta (Guangdong) in China or politically orchestrated economies such as Singapore and Taiwan, not only proved adept at adapting to financial pressures but were even able to create an oppositional force to demonstrate the vulnerability of finance capital—now heavily concentrated in the United States as well as Europe and Japan—to the production of real values. The fact that many of these industrial production complexes were regionally concentrated within a state, or even, in some instances, between states, is of considerable interest, for reasons that we will address in Chapter 3. Subtle lines of counter-attack against the hegemony of the United States in the realm of finance were emerging in the interstices of the worlds of production. And the sign of that was the piling up of trade balance surpluses, particularly in East and South-East Asia. The recycling of these surpluses back into the financial system made it seem, however, as if Wall Street was still the operative centre of the financial universe. While there had been, therefore, phases (such as that of the 1980s) when the hegemony of the United States was being openly questioned both internally and externally, by the end of the 1990s much of that doubt had dissipated. The security of the United States and its financial dominance in world affairs was assured. The boom in asset values within the United States and the rise of a 'new economy' built around supposedly strong productivity gains and a whole raft of dot.com companies kept the US economy growing rapidly enough to entrain the rest of the world into respectable rates of capital accumulation. Consumerism, the golden rule of internal peace within the United States, boomed to astonishing levels in the US as well as in the other centres of advanced capitalism.

This system has now run into serious difficulties. As in 1973–5, the causes are multiple, though this time the volatility and chaotic fragmentation of power conflicts within political-economic life make it hard to discern what is happening behind all the smoke and mirrors (particularly those of the financial sector). But in so far as the crisis of 1997–8 revealed that the main centre of surplus productive capacity lay in East and South-East Asia (and sought to visit devaluation singularly upon that region), the rapid recovery of some parts of East and South-East Asian capitalism (South Korea in particular) has forced the general problem of excess capacity (overaccumulation) back to the forefront of global affairs. The collapse of the much-celebrated 'new economy' in a rubble of failed dot.com companies in the United States, followed by accounting scandals that dramatically revealed that 'fictitious' capital could all too easily remain unredeemable, not only undermined the credibility of Wall Street but brought into question the relationship between finance capital and production. The threat of massive devaluation of capital loomed and, with the fall of asset values, there were tangible signs of that threat already being realized (most dramatically with respect to pension funds, which found it increasingly difficult to meet their obligations).

Either new arenas of profitable capital accumulation (such as China) must be opened up, or, failing that, there will have to be a new round of devaluation of capital. The question becomes: who will bear the brunt of a new round of that devaluation? Where will the axe fall? The trend towards 'regionalization' within the global economy then appears more worrying. Echoes of the geopolitical competition that became so destructive in the 1930s begin to be heard. US abandonment of the spirit if not the letter of the WTO rules against protectionism by the imposition of tariffs on steel imports in 2002 was a particularly ominous sign. The bursting of the speculative bubble revealed the vulnerability of the United States to devaluation. The gathering recession, evident early in 2001, after a decade or more of spectacular (even if 'irrational') exuberance and avid consumerism, gave further evidence of that vulnerability well before the jolt to the system administered by the events of 9/11. Was the golden rule of the incessant upward march of consumerism within the United States about to be broken?

A major faultline of instability lies in the rapid deterioration in the balance of payments situation of the United States. 'The same exploding imports that drove the world economy' during the 1990s, writes Brenner, 'brought US trade and current account deficits to record levels, leading to the historically unprecedented growth of liabilities to overseas owners' and 'the historically unprecedented vulnerability of the US economy to the flight of capital and a collapse of the dollar'.[23] But this vulnerability exists on both sides. If the US market collapses then the economies that look to that market as a sink for their excess productive capacity will go down with it. The alacrity with which the central bankers of countries like China, Japan, and Taiwan lend to cover US deficits has a strong element of self-interest: they thereby fund the US consumerism that forms the market for their products. They may now even find themselves funding the US war effort.

But the hegemony and dominance of the United States is, once more, under threat, and this time the danger seems more acute. Its roots lie in the unbalanced reliance upon finance capital as a means to assert hegemony. Historically, Arrighi (following Braudel) points out, financial expansions indicate 'not just the maturity of a particular stage of development of the capitalist world-economy, but also the beginning of a new stage'.[24] If financialization is a likely prelude to a transfer of dominant power from one hegemon to another (as has historically been the case) then the US turn towards financialization in the 1970s would appear to have been a peculiarly self-destructive move. The deficits (both internal and external) cannot continue to spiral out of control indefinitely, and the ability and willingness of others (primarily in Asia) to fund them is not inexhaustible. The sheer volume of support to the US is astonishing, rising to $2.3 billion a day at the beginning of 2003. Any other country in the world that exhibited such a macro-economic condition would by now have been subjected to ruthless austerity and structural adjustment procedures by the IMF. But the IMF is the United States. As Gowan remarks: 'Washington's capacity to manipulate the dollar price and to exploit Wall Street's international financial dominance enabled the US authorities to avoid doing what other states have had to do: watch the balance of payments; adjust the domestic economy to ensure high levels of domestic savings and investment; watch levels of public and private indebtedness; ensure an effective domestic system of financial intermediation to ensure the strong development of the domestic productive sector.' The US economy has had 'an escape route from all these tasks' and 'by all normal yardsticks of capitalist national accounting' has become 'deeply distorted and unstable' as a result.[25]

The power of the Wall Street–Treasury–IMF complex is both symbiotic with and parasitic upon a coercively imposed financial system built around the so-called Washington consensus and later elaborated through the construction of new international financial architecture. This, writes Soederberg, is clearly 'an annex of the US state', even though it also serves the interests of the 'transnational bourgeoisie as a whole'.[26] But the disciplining, even destruction, of the 'developmental' states centred in East and South-East Asia makes it tempting to bolt the system, much as Malaysia did when it suddenly, and quite successfully, abandoned the neo–liberal rules, refused the discipline of the IMF, and imposed capital controls of the sort that had not been seen since the 1960s. It is not clear how far this can go before regional alliances form and opt out, thus driving a stake through the heart of the Washington consensus and undermining the structure of the new financial architecture that has so far been so advantageous to the United States. Nor is it clear, as the tariff on steel imports shows, that the US will follow the rules. On this point it is worth recalling that the US

Senate ratification of the WTO agreement carried with it the proviso that the US could ignore and refuse any WTO ruling that it considered to be fundamentally unfair to US interests (a familiar stance in which the US assumes it has the right to be both judge and jury).

To cap it all, resistance towards and resentment of the powers of the Wall Street–Treasury–IMF complex are everywhere in evidence. A world-wide anti-globalization movement (quite different in form from the class struggles embedded in the processes of expanded reproduction) is morphing into an alternative globalization movement with a lot of grassroots support. Populist movements against US hegemony by formerly pliant subordinate powers, particularly in Asia (South Korea is a case in point) but also now in Latin America, threaten to transform grassroots resistance into a series of state-led if not intensely nationalist resistances to US hegemony. It is under these conditions that anti–imperialism begins to take on a different coloration which, in turn, helps define more clearly within the United States what its own imperialist project might have to be if it is to preserve its hegemonic position. If hegemony weakens, then the danger exists of a turn to far more coercive tactics of the sort we are now witnessing in Iraq.

Options

The options for the United States are limited. While Arrighi and his colleagues do not envisage any serious external challenge, they do worryingly conclude that the US has even greater capabilities than Britain did a century ago to convert its declining hegemony into exploitative domination. If the system eventually breaks down, it will be primarily because of US resistance to adjustment and accommodation. And conversely, US adjustment and accommodation to the rising economic power of the East Asian region is an essential condition for a non-catastrophic transition to a new world order.[27]

The Bush administration's shift towards unilateralism, towards coercion rather than consent, towards a much more overtly imperial vision, and towards reliance upon its unchallengeable military power, indicates a high-risk approach to sustaining US domination, almost certainly through military command over global oil resources. Since this is occurring in the midst of several signs of loss of dominance in the realms of production and now (though as yet less clearly) finance, the temptation to go for exploitative domination is strong. Whether or not this will lead later to a catastrophic break-up of the system (perhaps by a return to Lenin's scenario of violent competition between capitalist power blocs) is hard even to imagine let alone predict.

The US could, however, downgrade if not turn away from its imperialist trajectory by engaging in a massive redistribution of wealth within its borders and a redirection of capital flows into the production and renewal of physical and social infrastructures (dramatic improvements in public education and repair of patently failing infrastructures would be a good place to start). An industrial strategy to revitalize its still substantial manufacturing sector would also help. If it is to go very far, this strategy would also entail an internal reorganization of class power relations and transformative measures affecting social relations of a sort that the United States has refused systematically to contemplate ever since the Civil War. State-subsidized private consumerism would have to be replaced by projects oriented towards public well-being. But this would require even more deficit financing and/or higher taxation as well as heavy state direction, and this is precisely what the dominant class forces within the US adamantly refuse even to contemplate; any politician who proposes such a package will almost certainly be howled down by the capitalist press and their ideologists, and just as certainly lose any election in the face of overwhelming money power. Yet, ironically, a massive counter-attack within the US as well as within other core countries of capitalism (particularly in Europe) against the politics of neo-liberalism and the cutting of state and social expenditures might be one of the only ways to protect capitalism internally from its self-destructive and crisis-prone tendencies in the present conjuncture. A new 'new deal' is the very minimum, but it is by no means sure that this would really work in the face of the overwhelming excess capacity within the global system. It is salutary to remember the lessons of the 1930s: there is very little evidence that Roosevelt's 'New Deal' solved the problem of the Depression. It took the travails of war between capitalist states to bring territorial strategies back into line so as to put the economy back on a stable path of continuous and widespread capital accumulation.

Even more suicidal politically within the US would be to try to enforce by self-discipline the kind of austerity programme that the IMF typically visits on others. Any attempt by external powers to do so (by capital flight and collapse of the dollar, for example) would surely elicit a savage US political, economic, and even military response. It is hard to imagine that the US would peacefully accept and adapt to the phenomenal growth of East Asia and recognize, as Arrighi suggests it should, that we are in the midst of a major transition towards Asia as the hegemonic centre of global power. It is unlikely that the US will go quietly and peacefully into that goodnight. It would, in any case, entail a radical reorientation—some signs of which (as we will see in Chapter 3) already exist—of East Asian capitalism from dependency on the US market to the cultivation of an internal market within Asia itself. The gradual withdrawal of funds from the US would have calamitous consequences. But ever-expanding indebtedness is a perilous way to keep consumerism alive within the US, let alone pay for a war. The lesson of the crisis of 1973–5 was that at some point the capitalistic logic will come home to roost and expose the impossibility of a strategy of guns and butter for evermore.

It is in this context that we see the Bush administration looking to flex military muscle as the only clear absolute power it has left. The open talk of empire as a political option presumably seeks to hide the exaction of tribute from the rest of the world under a rhetoric of delivering peace and freedom for all. Control over oil supplies provides a convenient means to counter any power shifts—both economic and military—threatened within the global economy. The current situation reeks of a rerun of what happened in 1973, since Europe and Japan, as well as East and South-East Asia (now crucially including China), are even more heavily dependent on Gulf oil than is the United States. If the US successfully engineers the overthrow of both Chavez and Saddam, if it can stabilize or reform an armed-to-the-teeth Saudi regime that is currently based on the shifting sands of authoritarian rule (and in imminent danger of falling into the hands of radicalized Islam), if it can move on (as seems possible) from Iraq to Iran and consolidate its position in Turkey and Uzbekistan as a strategic presence in relation to Caspian Basin oil reserves (which the Chinese are desperately trying to buy into), then the US, through firm control of the global oil spigot, might hope to keep effective control over the global economy and secure its own dominance for the next fifty years. But much also depends, as Friedman noted in the passages cited in the Introduction, upon whether the US can persuade the world that it is acting in a leadership role, concerned to develop collective power by acting as guarantor of global oil supplies to all, or whether it is acting out of narrow self-interest to secure its own position at the expense of others. Is it, in short, resorting to domination through coercion or exercising leadership through hegemony? The most likely tactic is to try to mask the latter in a veneer of the former. But the failure to garner full international support for the invasion of Iraq suggests that much of the world is suspicious of US motivations.

The dangers of this strategy in the Gulf region are immense. Resistance will be formidable, not least from Europe and Asia, with Russia and China not far behind. The reluctance to sanction the US military invasion of Iraq in the UN, particularly by France, Russia, and China (who gained access to Iraqi oil exploitation during the 1990s), illustrates the point. And the internal dynamics of anti-American struggles in the Gulf region are as unpredictable as they are complex. The potentiality for destabilization of the whole region stretching from Pakistan to Egypt is considerable. The hubristic view that the whole structure of political power and territorial organization in the region—so arbitrarily created by the British and French as a side-bar to the Versailles agreements—can be remade and stabilized under the leadership of the US and its allies, is simply too far-fetched to contemplate (though there are strategists within the US government who seem to believe this is possible).

It is here, however, that the US is in the position to play its strongest card—military dominance—and to do so coercively if necessary. We know full well, from the defence planning documents issued over the last decade or so, what the political strategy is in this realm. It is to maintain military primacy at all costs and to discourage and resist the emergence of any rival superpower. The spread of weapons of mass destruction of any kind will be prevented, and the US should be prepared to use pre-emptive force if necessary to achieve that goal. During the Clinton years this was translated into an active capacity to fight two regional wars at the same time (and the examples chosen for planning purposes in 1995 were, interestingly, Iraq and North Korea). But the Cheney–Wolfowitz doctrine, first laid out in the last years of the former Bush administration and consolidated in the *Project for the New American Century* (which, interestingly, repeats Luce's move to disguise the territoriality of empire in the conceptual fog of a 'century') went further still. Fixed alliances (like NATO) are to be abandoned (they are too constraining) and ad hoc coalitions should be built on a case-by-case basis. In this way the US would no longer be bound by the views of its allies. The US reserves the right to go it alone if necessary with overwhelming military firepower. It overtly claims the right of pre-emptive strike to head off nuclear, biological, or chemical attacks, to protect access to key strategic raw materials (such as oil), and protect against terrorist attacks or other threats (such as economic strangulation). What is so interesting about these defence strategy documents from 1991–2 is how closely their prescriptions are now being followed. Armstrong, after a close study of these documents concludes:

The Plan is for the United States to rule the world. The overt theme is unilateralism, but it is ultimately a story of domination. It calls for the United States to maintain its overwhelming military superiority and prevent new rivals from rising up to challenge it on the world stage. It calls for dominion over friends and enemies alike. It says not that the United States must be more powerful, or most powerful, but that it must be absolutely powerful.[28]

The irony in all this, as Armstrong goes on to note, is that, having helped bring down the Soviet Union, the US is now pursuing the very politics for which that 'evil empire' was condemned and opposed. The US ought, as Colin Powell graphically puts it, 'to be the bully on the block'. The rest of the world would happily accept this, he went on to assert confidently, because the US 'can be trusted not to abuse that power'.

There is, in this, another possible irony: if the Soviet empire was really brought down by excessive strain on its economy through the arms race, then will the US, in its blind pursuit of military dominance, undermine the economic foundations of its own power? Regional military commitments are enormous and growing. The US was already spending $4–5 billion a year on patrolling the Gulf region before the military build-up began. Already the Bush administration has requested nearly $75 billion for the war, and that is only until September 2003. The total cost is unlikely to be less than $200 billion, according to plausible estimates, and this presumes no unintended disaster, such as regional break-up and extensive civil war. And the US plans 'normal' spending on its military that is equal to that of the rest of the world. The danger of over-reach is serious, particularly since federal budget deficits loom larger and larger in the fiscal landscape and budget crises at the state and local levels are already biting hard into levels of public service provision. It will then be doubtful if the golden rule that has prevailed since Roosevelt—that expenditures on imperial purposes abroad should not interfere with the endless spiral of consumerism at home—can be maintained. The US will not merely have to sacrifice precious blood for oil and the maintenance of an ailing hegemony; it may have to sacrifice its whole way of life too. The capitalistic logic of power will tear the territorial logic that is now being pursued to shreds.

Regional and Counter Hegemons

The triadic regional structure within the global economy, with North America supposedly at the apex, is not necessarily a stable configuration. The formal arrangements set up within the European Union appear to offer the possibility of an integrated European economy at least as large and as powerful as that of the United States. At the very least this presages the formation of a regional hegemon and perhaps the emergence of a real competitor with the United States.[29] The capitalist logic within the EU, though by no means spectacular, seems to be working well enough. Interlinkages and networked relations within the economy are both proliferating and consolidating across the European space. The transition to a single currency was achieved relatively painlessly, and the potential for the euro to challenge the dollar as the reserve currency of choice, though muted, is nevertheless real (Saddam's proposal to denominate his oil sales in euros rather than dollars may well be another significant reason for the US to insist upon regime change rather than disarmament in Iraq). But the EU is politically fragmented and its overall territorial logic remains indeterminate. The US has all manner of levers to divide and rule and thereby frustrate the emergence of any clear territorial logic at the European level. It seeks to prevent the emergence of a 'fortress Europe' by a double strategy of (*a*) insisting upon the rules of neo-liberalism as the basis for exchange relations and capital flows (hence the importance of the WTO) and (*b*) keeping certain political and military levers in place whereby it can influence internal politics of the EU. This entails engaging with individual European states on a bilateral basis rather than with Europe as a whole, and cultivating special alliances (e.g. with Britain, Spain, and Italy as well as with the tier of eastern bloc countries, with Poland at the centre, that are poised for admission). Though the US itself now proposes to abandon fixed alliances, it still hangs on to

NATO in spite of its general irrelevance given the end of the Cold War, in part because it keeps European military planning and development under US command. The US supports, for example, the idea that Europe should develop its own military rapid-response force but only on the condition that it remain under NATO command. The fact that NATO does not correspond to the EU is, for the US, a distinct advantage since it makes it even more difficult to render the territorial logic into a coherent political and military force.

Divisions within the EU, mainly between pro-American countries and those seeking to assert an independent politics, are at this point too severe to imagine a common foreign and military strategy. It is unlikely that the EU will produce a coherent basis for its own 'territorial logic of power' to be projected upon the world in the very near future. But things on that front can change quickly, particularly if the US administration continues to approach European opinion with such a withering mix of contempt and callous disregard. The EU certainly constitutes a regional hegemon, but its potentiality to rival the US is currently confined to the spheres of production and finance.

At this point in time, the challenge to US dominance posed by East and South-East Asia seems far more serious. Financial and productive power have continued to accumulate in the region, draining power away from North America as well as, to a lesser degree, from Europe. Unlike Europe, the region shows little sign of any attempt to create a formal structure of political-military power, and the relationships between states are networked rather than formal, capitalistic rather than territorial. The United States in any case currently exercises a level of political and military control over the governments of Japan, Taiwan, and, until very recently, South Korea which would make any independent political moves by these countries difficult. While it seems unlikely, therefore, that any coherent territorial logic of power will develop in the region, the power of the capitalistic logic looks more and more overwhelming and prospectively hegemonic in the global economy, particularly as the huge weight of China and, to a lesser degree, India increasingly enter into the balance. We will take up the economic consequences of these shifts in Chapter 3, but a political and military question does arise, because China is not dominated by the United States in the same way as Japan and has the capacity and, at times it seems, the willingness to take on a territorial leadership role within the region as a whole. The political and military containment of China would be just as essential to the maintenance of US global hegemony as would be a politics of divide and rule for Europe. And in this, as I observed in Chapter 1, the control over Middle Eastern oil reserves would serve US interests very well if it ever felt it necessary to rein in Chinese geopolitical ambitions. In all of this, however, there is a delicate balance between keeping the world open enough to allow the capitalistic logic to unfold relatively free of constraints and keeping territorial logics stable and confined enough to prevent the rise of any grand challenge to US military and political dominance.

But these are not the only configurations of territorial power that can be imagined. While the relative fixity of territorial arrangements militates against fluidity, rapid shifts in the nature of alliances can and do occur. When, for example, US policy towards Iraq created a bond of resistance early in 2003 between France, Germany, and Russia, even backed by China, it became possible to discern the faint outlines of a Eurasian power bloc that Halford Mackinder long ago predicted could easily dominate the world geopolitically. That the US had long been nervous of such a power bloc was evident in the way it responded strongly to de Gaulle's overtures to the Soviet Union in the 1960s and to Willy Brandt's 'Ostpolitik' of the 1970s. That the US still has much to fear from such an alignment was forcibly expressed by Henry Kissinger when he remarked that this new alignment presaged a return to a balance of power politics typical of the nineteenth century, ruefully adding that in this case 'it is not evident that the US will lose', thus admitting the very real possibility that it might.[30] The fact that the Bush administration could bring about such a fearsome counter-alliance in the space of less than a year illustrates how fast geopolitical realignments can occur and how easily catastrophic mistakes can unravel years of careful cultivation of diplomatic and military protections. The US invasion of Iraq then takes on an even broader meaning. Not only does it constitute an attempt to control the global oil spigot and hence the global economy through domination over the Middle East. It also constitutes a powerful US military bridgehead on the Eurasian land mass which, when taken together with its gathering alliances from Poland down through the Balkans, yields it a powerful geostrategic position in Eurasia with at least the potentiality to disrupt any consolidation of a Eurasian power that could indeed be the next step in that endless accumulation of political power that must always accompany the equally endless accumulation of capital.

The end of the Cold War clearly implied that big changes were on the way. The territorial logics of power are in the course of mutation, but the outcomes are by no means certain. It is now also evident that the territorial and the capitalistic logics exist in a state of high tension. Under Bush, the US territorial logic has been made clear, which is why all the current talk of empire and the new imperialism is so US-centred. But the balance of forces at work within the capitalistic logic point in rather different directions. How this will all turn out depends mightily, therefore, upon a better understanding of how the capitalistic logic of power is working. It is this question that will be taken up in Chapter 3.

3

Capital Bondage

The survival of capitalism for so long in the face of multiple crises and reorganizations accompanied by dire pre-
dictions, from both the left and the right, of its imminent
demise, is a mystery that requires illumination. Lefebvre,
for one, thought he had found the key in his celebrated
comment that capitalism survives through the production
of space, but he unfortunately failed to explain exactly
how or why this might be the case.[1] Certainly both Lenin
and Luxemburg, though for quite different reasons and
utilizing quite different forms of argument, considered
that imperialism—a certain form of production and utilization of the global space—was the answer to the riddle,
though in both cases this solution was finite and therefore
replete with its own terminal contradictions.

It was in this context that, in a series of publications
beginning more than twenty years ago, I proposed a
theory of a 'spatial fix' (more accurately a spatio-temporal
fix) to the crisis-prone inner contradictions of capital
accumulation.[2] The central point of this argument
concerned a chronic tendency within capitalism, theoretically derived out of a reformulation of Marx's theory of

the tendency for the profit rate to fall, to produce crises of overaccumulation.[3] Such crises are typically registered as surpluses of capital (in commodity, money, or productive capacity forms) and surpluses of labour power side by side, without there apparently being any means to bring them together profitably to accomplish socially useful tasks. The most obvious case of this was the world-wide slump of the 1930s when capacity utilization was at an all-time low, surplus commodities could not be sold, and unemployment was at an all-time high. The effect was to devalue and in some cases even destroy the surpluses of capital and to reduce the surpluses of labour power to a miserable state. Since it is the lack of profitable opportunities that lies at the heart of the difficulty, the key economic (as opposed to social and political) problem lies with capital. If devaluation is to be avoided, then profitable ways must be found to absorb the capital surpluses. Geographical expansion and spatial reorganization provide one such option. But this option cannot be divorced from temporal shifts in which surplus capital gets displaced into long-term projects that take many years to return their value to circulation through the productive activity they support. Since geographical expansion often entails investment in long-lived physical and social infrastructures (in transport and communications networks and education and research for example), the production and reconfiguration of space relations provides one potent way to stave off, if not resolve, the tendency towards crisis formation under capitalism. The US government tried to respond to the overaccumulation problem in the 1930s, for example, by setting up future-oriented public works projects in hitherto undeveloped locations with the direct intention of mopping up the surpluses of capital and labour then available (it was in the same spirit, incidentally, that the Nazis built the autobahns during these years).

The capitalistic (as opposed to territorial) logic of imperialism has, I argue, to be understood against this background of seeking out 'spatio-temporal fixes' to the capital surplus problem (and it is, I repeat, the capital surplus rather than the labour surplus that must be the primary focus of analytic attention). In order to understand how this happens, I must first describe, albeit in schematic and very general terms, how capital circulates in space and time to create its own distinctive historical geography. In so doing, I will try to keep the dialectical relationship between the politics of state and empire on the one hand and the molecular movements of capital accumulation in space and time on the other, firmly at the centre of the argument. I therefore begin with some basic observations on the importance of the state as a territorialized framework within which the molecular processes of capital accumulation operate.

State Powers and Capital Accumulation

Capital accumulation through price-fixing market exchange flourishes best in the midst of certain institutional structures of law, private property, contract, and security of the money form. A strong state armed with police powers and a monopoly over the means of violence can guarantee such an institutional framework and back it up with definite constitutional arrangements. State

formation, coupled with the emergence of bourgeois constitutionality, have therefore been crucial features within
the long historical geography of capitalism.

Capitalists do not absolutely require such a framework
to function, but without it they do face greater risks. They
have to protect themselves in environments that may not
recognize or accept their rules and ways of doing business.
Merchants and dealers can survive by setting up their own
codes of honour and of action (much as the street money
traders still do throughout much of the Middle East).
They develop networks of trust among themselves (sometimes relying on family—as did the Rothschilds in the
nineteenth century—and kinship) and substitute their
own violence (as merchant capitalists have often done)
either within or against state power to protect their property and business activities from the threat of antagonistic
forces or state powers. They may need to go against state
law where state powers are either hostile (as was the case
in many formerly communist countries) or indifferent to
their activities.[4] This lawlessness can take on perverse
forms with mafias, drug cartels, and the like, even in the
heart of strong pro–capitalist states. In other instances,
capitalists can secure protected enclaves for themselves.
The town charters of medieval Europe created islands of
bourgeois citizens right in the midst of feudal relations.
The East India or the Hudson Bay companies' trading
posts, and the enterprise zones for foreign investment now
set up in, say, China, are other examples. The molecular
processes of capital accumulation can and do create their
own networks and frameworks of operation over space in
innumerable ways, using kinship, diasporas, religious and
ethnic bonding, and linguistic codes as means to produce

intricate spatial networks of capitalist activity independent of the frameworks of state power.

Nevertheless, the preferred condition for capitalist activity is a bourgeois state in which market institutions and rules of contract (including those of labour) are legally guaranteed, and where frameworks of regulation are constructed to contain class conflicts and to arbitrate between the claims of different factions of capital (for example between merchant, finance, manufacturing, agrarian, and rentier interests). Policies with respect to security of the money supply and towards foreign trade and external affairs must also be structured to advantage business activity.

Not all states act in an appropriate way, of course, and even when they do they exhibit a variety of institutional arrangements that can produce quite different results. Much has therefore depended on how the state has been constituted and by whom, and what the state was and is able or prepared to do in support of or in opposition to processes of capital accumulation. The state, as we will see in Chapter 4, played a key role in original or primitive accumulation, using its powers not only to force the adoption of capitalistic institutional arrangements but also to acquire and privatize assets as the original basis for capital accumulation (the appropriation of Church property in the Reformation or the enclosure of common lands through state action in Britain being obvious examples). But the state also takes on all manner of other influential roles (taxation being one). Differences in state formation and in state policies have always been important. The British state, being influenced far more by merchant capitalists, played a quite different role in relation to accumulation to France, where landed interests predominated. The two countries even produced quite different economic theories to explain and justify their stances. The British became attached to the mercantilism of Munn's *England's Treasure by Foreign Trade*, which focused on the accumulation of bullion out of trade, while the French supported the physiocratic notion that all wealth (value) derived from the land and that trade and industry were therefore secondary and parasitic forms of wealth creation. State power hostile to private accumulation of wealth—as has long been the case until very recently in China—can hold a country back. Social democratic states typically seek to curb excessive exploitation of labour power and place themselves behind the class interests of labour without abolishing capital. On the other hand, the state can be an active agent of capital accumulation. The developmental states of East and South-East Asia (like Singapore, Taiwan, and South Korea) have themselves affected the dynamics of capital accumulation directly through their actions (often by holding down the aspirations of labour). But then this kind of state interventionism has long existed. Bismarck's Germany and Meiji restoration Japan rose to prominence as territories of capital accumulation in part because of the strong supportive if not forcing role of state power. And the *dirigiste* tradition in France (as exemplified in the Gaullist policies of the 1960s) gave a definite quality to accumulation that differentiated it from, say, Britain (as everyone recognizes as they travel the rail systems). And, of course, when it comes to struggles over hegemony, colonialism, and imperial politics, as well as over more mundane aspects of foreign relations, the state has long been and continues to be the fundamental agent in the dynamics of global capitalism.

States are not the only relevant territorial actors. Collections of states (regional power blocs that may either be informally networked as in East and South-East Asia or more formally constituted as in the European Union) cannot be ignored, any more than can sub-national entities, such as regional governments (states in the USA) and metropolitan regions (Barcelona plus Catalonia, or the San Francisco Bay area). Political power, territorialized governance, and administration are constituted on a variety of geographical scales and constitute a hierarchically ordered set of politically charged environments within which the molecular processes of capital accumulation occur.

But to depict the evolution of capitalism as an expression merely of state powers within an inter-state system characterized by competitive struggles for position and hegemony—as tends to happen in much of world systems theory—is far too limiting. It is just as erroneous as depicting the historical–geographical evolution of capitalism as if it were totally unaffected by territorial logics of power. But Arrighi raises an important question: how does the relative fixity and distinctive logic of territorial power fit with the fluid dynamics of capital accumulation in space and time?[5] To answer that we need first to specify how the molecular processes of capital accumulation actually work in space and time. In so doing I shall for convenience presume the prior existence of an appropriate and stable set of institutional arrangements guaranteed and facilitated by state power.

The Production of a Space Economy

In a number of earlier publications I set out a detailed theory of how a space economy emerges out of processes of capital accumulation.[6] I here take up the salient points of this argument only in summary form.

Exchanges of goods and services (including labour power) almost always entail changes of location. They define, at the very outset, an intersecting set of spatial movements that create a distinctive geography of human interaction. These spatial movements are constrained by the friction of distance and therefore the trace they leave upon the land invariably records the effects of such friction, more often than not causing activities to cluster together in space in ways that minimize such frictions. Territorial and spatial divisions of labour (the distinction between town and country being one of its most obvious early forms) arise out of these interacting exchange processes over space. Capitalist activity thereby produces uneven geographical development, even in the absence of the geographical differentiation in resource endowments and physical possibilities that add their weight to the logic of regional and spatial differentiations and specializations. Driven by competition, individual capitalists seek competitive advantages within this spatial structure and therefore tend to be drawn or impelled to move to those locations where costs are lower or profit rates higher. Surplus capital in one place can find employment somewhere else where profitable opportunities have not yet been exhausted. Locational advantages play a role for individual capitalists similar to those derived from technological advantages, and in certain situations the one may substitute for the other.

Capital Bondage

In some respects this line of argument parallels that of classical location theory (as laid out in the works of von Thünen, Alfred Weber, and Lösch and later synthesized in the work of Isard).[7] The main difference lies in the fact that those works typically sought to identify a spatial equilibrium in the geographical landscape of capitalistic activity, whereas in this case the processes of capital accumulation are seen as perpetually expansionary and therefore permanently disruptive of any tendency towards equilibrium. Classical location theory, furthermore, assumed an economic rationality that has little to do with actual capitalistic behaviours. For example, it defined what it called 'the spatial range of a good' in terms of the radial distance from a point of production where the market price (measured as production plus transport cost) placed it beyond what consumers would be willing or able to pay for it. But goods do not take themselves to market, merchants do. The historical role of merchant capitalists has entailed the constant probing and rolling back of spatial barriers (often well beyond that which would be considered 'rational') and the opening up of new modalities of movement and spaces for trade. Faced, for example, with confined local markets and high transport costs, medieval merchants became itinerant pedlars who sold their wares on the move over vast areas. In exactly the same way that competitive behaviour forces strong impulses of disruptive technological dynamism into capitalist economies (as individual capitalists seek competitive advantage by adopting a superior technology) so it also generates a state of perpetual motion and chronic instability in the spatial distribution of capitalistic activities as capitalists search for superior (i.e. lower-cost) locations. The geographical landscape of capitalist production, exchange, distribution, and consumption is never in equilibrium.

The competition within a spatial system is, however, as the neoclassical theorists of spatial order (Chamberlain, Hotelling, and Lösch) correctly recognized, a species of *monopolistic* competition.[8] This strange hybrid form of competition arises in the first instance because of the exclusions that derive from uniqueness of location. Spatial location always confers a certain monopolistic advantage. Private property in land entails at its very basis a certain monopolistic power: no one can place their factory where my factory is already located. And if very special advantages attach to my location, then those advantages belong to me alone. This allows free play within a space economy to the capitalistic preference for monopoly control as opposed to open competition. Though the abstract theory of capitalism (including its neo-liberal variant) appeals all the time to the ideals of competition, capitalists covet monopoly powers because they confer security, calculability, and a generally more peaceful existence. Furthermore, the end product of competition is monopoly or oligopoly and the fiercer the competition the faster the system converges upon such states: witness the incredible rise in oligopoly and mono-poly situations in many sectors of the economy (from air-lines and energy to the media and entertainment) during the last thirty years of neo-liberal hegemony in economic policy in the core capitalist states. Capitalists can and do use spatial strategies to create and protect monopoly powers wherever and whenever they can. Control over key strategic locations or resource complexes is an impor-tant weapon. In some instances monopoly power becomes strong enough to inhibit the dynamism in capitalism's geography, introducing strong tendencies towards geographical inertia and stagnation. The tendency towards spatial dynamism given by the competitive search for profits is countered by the bundling together of monopoly powers in space. It is from exactly such centres that imperialist practices and calls for an imperial presence in the world typically emanate. Lenin and Hilferding were therefore right to emphasize the important inner connection between monopolization and imperialism.

The asymmetries in exchange identified in Chapter 2 as crucial to understanding the economic logic of imperialism arise out of monopolistic competition. The resultant inequalities take on a specific spatial and geographical expression, usually as concentrations of privileges and powers in certain places rather than in others. In the past, high transport costs and other barriers to movement (such as tariffs, tolls, and quotas) meant the existence of many local monopolies. I ate local food and drank local beer because the high friction of distance gave me no other choice. Protections of this sort break down, however, as transport costs diminish and as political barriers to trade are removed through arrangements such as the WTO. I eat vegetables from California in Paris and drink imported beers from all over the world in Pittsburgh. Even Detroit automakers, who in the 1960s were considered an exemplar of the sort of oligopoly condition characteristic of what Baran and Sweezy defined as 'monopoly capitalism',[9] found themselves seriously challenged by foreign, particularly Japanese, imports. Capitalists have therefore had to find other ways to construct and preserve their much-coveted monopoly powers. The two major moves they have made is towards massive centralization of capital, which seeks dominance through financial power, economies of scale, and market position, and avid protection of technological advantages (always, as I have already pointed out, a substitute for locational advantages) through patent rights, licensing laws, and intellectual property rights. It is no accident that the latter has been the focus of intense negotiation within the WTO, producing the so-called TRIPS (trade-related intellectual property rights) agreement.

All of this points up how important is the ability to move commodities, productive capacity, people, and money over space. The conditions prevailing within the transport and communications industries are key here. Throughout capitalist history, technological innovations within this field have dramatically altered the conditions of spatiality (the friction of distance) and generated all manner of instabilities within the space economy of capitalism. The reasons behind the tendency towards what Marx called 'the annihilation of space through time' have been laid out at length elsewhere and I see no point in repeating them here.[10] But what can be derived theoretically, and which jibes with capitalism's historical-geographical record, is an incessant drive towards the reduction if not elimination of spatial barriers, coupled with equally incessant impulses towards acceleration in the turnover of capital. The reduction in the cost and time of movement has proven a compelling necessity of a capitalist mode of production. The trend towards 'globalization' is inherent in this, and the evolution of the geographical landscape of capitalist activity is driven remorselessly by round after round of time–space compression.

Capital Bondage

One of the further consequences of this process is a perpetual impulse towards the transformation of the geographical scale at which capitalist activity gets defined. Just as the coming of the railways and the telegraph in the nineteenth century completely reorganized the scale and diversity of regional specializations, and of urbanization and 'regionality' more generally, so the more recent round of innovations (everything from jet transport and containerization to the internet) has changed the scale at which economic activity is articulated. Without these impulses, the changing scale of hegemonic power, which was noted in Chapter 2, would be both materially impossible and theoretically incomprehensible. Political re-territorializations such as the European Union (dreamed of during the Enlightenment and actively proposed by utopian thinkers such as Saint-Simon in the early nineteenth century) become not only more practicable but more and more of an economic necessity. This is not to say, of course, that political shifts are simply a function of these material transformations in space relations; matters are far more complicated than that. But changing space relations do function as necessary conditions shaping the political reorganizations we see around us. Here, as we shall shortly see, is one crucial point where the territorial and capitalist logics of power intersect.

The particular conditions in the transport and communications industry illustrate a more general problem. Fluid movement *over* space can be achieved only by fixing certain physical infrastructures *in* space. Railways, roads, airports, port facilities, cable networks, fibre-optic systems, electricity grids, water and sewage systems, pipelines, etc., constitute 'fixed capital embedded in the land' (as opposed to those forms of fixed capital, such as aircraft and machinery, that can be moved around). Such physical infrastructures absorb a lot of capital, the recovery of which depends upon their use *in situ*. Capital invested in a port facility to which no ships come will be lost. While fixed capital invested in the land facilitates spatial mobility for other forms of capital and labour, it demands that spatial interactions follow the fixed geographical patterning of its investments in order for its own value to be realized. The effect is for fixed capital embedded in the land—and this includes factories, offices, housing, hospitals, and schools as well as the capital embedded in transport and communications infrastructures—to act as a significant drag upon geographical transformations and the relocation of capitalist activity. Once again, we discover forces making for geographical inertia as opposed to dynamism. The capital locked into the physical infrastructures of New York City, London, or Tokyo–Yokohama is substantial and, as the brief interruption that occurred in New York around 9/11 showed so clearly, any interruption of the flows of capital into and through such locations can have a catastrophic economic effect. Furthermore, the distinctive patterning of these investments opens up more ways in which the monopolistic privileges that attach to location can be captured by individual capitalists. The developer who just happens to control the land where a major highway intersection is projected can make a speculative killing on the value of the land, as well as on the investments (such as office blocks and hotels) placed upon it.

It should be evident from the narrative so far that the geographical landscape of capitalist activity is riddled with contradictions and tensions and that it is perpetually unstable in the face of all manner of technical and economic pressures operating upon it. The tensions between competition and monopoly, between concentration and dispersal, between centralization and decentralization, between fixity and motion, between dynamism and inertia, between different scales of activity, all arise out of the molecular processes of endless capital accumulation in space and time. And these tensions are caught up in the general expansionary logic of a capitalist system in which the endless accumulation of capital and the never-ending search for profits dominates. The aggregate effect is, as I have often had cause to formulate it in the past, that capitalism perpetually seeks to create a geographical landscape to facilitate its activities at one point in time only to have to destroy it and build a wholly different landscape at a later point in time to accommodate its perpetual thirst for endless capital accumulation. Thus is the history of creative destruction written into the landscape of the actual historical geography of capital accumulation.

Political/Territorial versus Capitalist Logics of Power

The molecular processes of capital accumulation operating in space and time generate passive revolutions in the geographical patterning of capital accumulation. But the tensions and contradictions I have identified can also produce geographical configurations that achieve stability, at least for a time. I shall refer to these relatively stable configurations as 'regions', by which I mean regional economies that achieve a certain degree of structured coherence to production, distribution, exchange, and consumption, at least for a time. The molecular processes converge, as it were, on the production of 'regionality'. This is not, of course, a unique finding. It is very familiar territory to many historical and economic geographers, as well as to economic historians like Sydney Pollard, who emphasize regional development and the development of regions as a fundamental feature in British economic development. There is a long tradition in economic theory, from Alfred Marshall (with his emphasis upon industrial—now called 'Marshallian'—production districts) through Perroux (with his emphasis upon growth poles) to Paul Krugman (with his interest in 'self-organizing' regional economies), that sees the production of regional organization as both an inevitable consequence and a basic condition for understanding the dynamics of capital accumulation.[11] Political scientists such as Mittelman have recently emphasized the importance of regional organization at both the supra- and sub-national levels in understanding the complex cross-currents at work within the global economy.[12]

The boundaries of regions of this sort are always fuzzy and porous, yet the interlocking flows within the territory produce enough structured coherence to mark the geographical area off as somehow distinctive relative to all other areas within a national economy or beyond. Structured coherence usually extends well beyond pure economic exchanges, fundamental though these may be, for it typically encompasses attitudes, cultural values, beliefs, and even religious and political affiliations among both capitalists and those whom they employ. The necessity to produce and maintain collective goods requires that some system of governance be brought into existence and preferably formalized into systems of administration within the region. Dominant classes and hegemonic class alliances can form within the region and lend a specific character to political as well as to economic activity. They have to be concerned about public goods, and may therefore find themselves forced to engage in public provision. The formation of physical and social infrastructures both to support economic activity but also to secure and promulgate cultural and educational values and many other aspects of civic life typically reinforces the coherence of what begins to emerge as a regional entity within the global economy. Patterns of trade and competition, and specialization and concentration on key industries or technological mixes or on particular labour relations and skills, interlink regional economies loosely into some patterned whole of uneven geographical development. What exactly happens with respect to internal dynamics and external relations depends on the class structure that arises and the forms of class alliance that form in and around the issues of governance.[13]

The fundamental point to recognize, however, is that a certain informal, porous but nevertheless identifiable territorial logic of power—'regionality'—necessarily and unavoidably arises out of the molecular processes of capital accumulation in space and time, and that interregional competition and specialization in and among these regional economies consequently becomes a fundamental feature of how capitalism works. This then poses the key question: how does this evolving regionality arrived at through the molecular processes of capital accumulation operating in space and time correlate with the territorial logic of power as expressed through the politics of state and empire?

The answer in the first instance is that they have nothing necessarily to do directly with one another. Pollard, for example, estimates that the regional economies that played such a key role in Britain's industrial revolution in the closing years of the eighteenth century were no more than twenty miles across, effectively small islands in a much grander British polity whose boundaries had been fixed upon at least two hundred years before.[14] But these small islands created impulses that were eventually to engulf the whole nation. As time went on and transport and communications systems changed, so these small islands grew and merged into much larger regions taking over, for example, Birmingham and the whole of the Midlands, Manchester and the whole of southern Lancashire and the West Yorkshire conurbation. So influential did these regions become that their politics and interests came to play a very influential, if not determining, role in how the nation as a whole was governed. They even spawned their own particular philosophies, with the 'Manchester school' of free traders, led by Cobden and Bright, daring to dress up their special interests as those of the nation as a whole. Birmingham, as personified in the figure of 'Radical Joe' Chamberlain, took a rather different view, as we shall see. It is nevertheless fair to say that the politics of state for Britain as a whole were captured by regional interests which were not necessarily those of the rest of the country (even poor Scotland rarely got a look in). The axis that runs from London through Birmingham and the Midlands and up to the conurbations of Lancashire and Yorkshire dominated British politics for the best part of a century and still exerts enormous pull and power. This same sort of tale could be told across Europe, and of course region and section in the United States have been of very great importance historically, as power has shifted from the north-east and midwest to the south, south-west, and the Pacific Rim.[15] The Pearl River delta and lower Yangtze (Shanghai) encapsulate dynamic power centres within China that economically (though not necessarily politically) dominate the rest of the country. The container that is the territorial state is, in short, often captured by some dominant regional interest or coalition of interests within it, until, that is, some other region arises to counter or supersede it. These shifts of influence from one region to another, from one scale to another, are precisely what the passive revolutions deriving from the molecular processes of endless capital accumulation typically accomplish. But the general principle is clear: regionality crystallizes according to its own logic out of the molecular processes of capital accumulation in space and time. In due course the regions thus formed come to play a crucial role in how the body politic of the state as a whole, defined solely according to some territorial logic, positions itself.

But the state is not innocent, nor is it necessarily passive, in relation to these processes. Once it recognizes the importance of fostering and capturing regional dynamics as a source of its own power it can seek to influence those dynamics by its policies and actions. It may in the first instance do so accidentally. In the nineteenth century, for example, states built roads and communications systems primarily for purposes of administration, military control, and protection of the territory as a whole. But, once built, these infrastructures provided paths that more easily facilitated the flow of goods, labour, and capital. In many instances the investments were jointly conceived. It is still a matter of debate as to whether Haussmann built the new boulevards of Paris after 1853 primarily for purposes of military control over a restive population or as a means to facilitate the easier circulation of capital within the confines of a city straitjacketed in a medieval network of streets and alleys.[16] And, interestingly, while the inter-state highway system of the United States was almost certainly built primarily for economic reasons, its legitimacy was pressed on the public in the name of national security and defence.

But the state can use its powers to orchestrate regional differentiation and dynamics not only through its command over infrastructural investments (particularly in transport and communications, education, and research) but also through its own imposition of planning laws and administrative apparatuses. Its powers to accomplish reforms of the basic institutions necessary for capital accumulation can also have profound effects (both positive and negative). When, for example, local banking was supplanted by national banks in Britain and France in the nineteenth century, the free flow of money capital across the national space altered regional dynamics. More recently, the abolition in the United States of restrictive local banking laws, followed by a wave of takeovers and mergers of regional banks, has changed the whole investment climate in the country away from local and into a more open and fluid construction of regional configurations. And in certain instances, Singapore being the most exemplary case, a political state can actually set out to build an effective and dynamic regional economy within itself by systematically capturing the molecular processes of capital accumulation in space and time within its borders. As is now well known, an attractive business climate is likely to be a magnet for capital flow, and so states go out of their way to augment their own powers by setting up havens for capital investment. In so doing they are using, as always, the monopoly powers inherent in space to try to offer monopoly privileges to whoever can take advantage of them.

This leaves us with the final problem of what happens when the molecular processes of region construction overflow the borders of the political state or for some reason require an outlet beyond those borders. There are, of course, some fascinating cases of regional economies that straddle national boundaries—El Paso and Ciudad Juarez or Detroit and Windsor are interesting examples. And the formation of supra-state administrative structures such as the European Union, or even just a common market such as NAFTA (North American Free Trade Agreement) or MERCOSUR (the common market of the Southern cone countries of Latin America), may be seen as solutions to this problem. But the really big issue is what happens to surplus capitals generated within sub-national regional economies when they cannot find profitable employment anywhere within the state. This is, of course, the heart of the problem that generates pressures for imperialist practices in the inter-state system.

The evident corollary of all this is that geopolitical conflicts would almost certainly arise out of the molecular processes of capital accumulation no matter what the state powers thought they were about, that these molecular movements (particularly of finance capital) can easily undermine state powers, and that the political state, in advanced capitalism, has to spend a good deal of effort and consideration on how to manage the molecular flows to its own advantage both internally and externally. And on the external front it will typically pay great attention to those asymmetries that always arise out of spatial exchanges and attempt to play the cards of monopoly control as strongly as it can. It will, in short, necessarily engage in geopolitical struggle and resort, when it can, to imperialist practices. We will see more concretely how this works in what follows.

The Circuits of Capital

The preceding analysis of spatio-temporal dynamics, though it pays due attention to general contradictions and instabilities, ignores the pervasive tendency of capitalism to produce crises of overaccumulation. We now need to examine more closely how the general processes of production of space become caught up in processes of crisis formation and resolution. Since it will be useful to refer to empirical examples in what follows, I propose to accept the empirical evidence offered by Brenner, which sees a chronic and enduring problem of overaccumulation pervading the whole of capitalism since the 1970s.[17] This will set the stage for interpreting the volatility of international capitalism since that time as a series of temporary spatio-temporal fixes that failed even in the medium run to deal with problems of overaccumulation.

Capital Bondage

The basic idea of the spatio-temporal fix is simple enough. Overaccumulation within a given territorial system means a condition of surpluses of labour (rising unemployment) and surpluses of capital (registered as a glut of commodities on the market that cannot be disposed of without a loss, as idle productive capacity and/or as surpluses of money capital lacking outlets for productive and profitable investment). Such surpluses can be potentially absorbed by (*a*) temporal displacement through investment in long-term capital projects or social expenditures (such as education and research) that defer the re-entry of capital values into circulation into the future, (*b*) spatial displacements through opening up new markets, new production capacities, and new resource, social, and labour possibilities elsewhere, or (*c*) some combination of (*a*) and (*b*).

The most interesting case is the combination of (*a*) and (*b*), but I first take up the solely temporal version which is illustrated in Figure 1. Flows of capital are drawn off from the realm of immediate production and consumption (the primary circuit) and redirected into either a secondary circuit of fixed capital and consumption fund formation or into a tertiary circuit of social expenditures and research and development. The secondary and tertiary circuits absorb excess capital into investments of long duration. Within the secondary circuit of capital, flows divide into fixed capital for production (plant and equipment, power-generating capacity, rail links, ports, etc.) and the creation of a consumption fund (housing, for example). Joint uses are often possible (the highway can be used for both production and consumption activities). A portion of the capital flowing into the secondary circuit

Capital Bondage

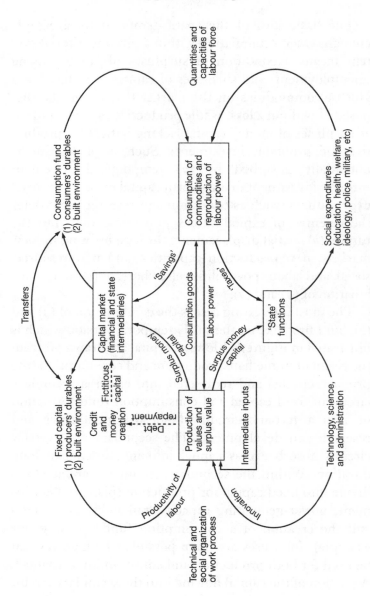

Fig 1. The Paths of Capital Circulation

is embedded in the land and forms a bank of physical assets in place—a built environment for production and consumption (everything from industrial parks, ports and airports, transport and communications nets to sewage and water systems, housing, hospitals, schools). These investments typically form a physical core to what a region is all about. They play, in short, a fundamental role in the production of regionality. Plainly, they constitute far more than a minor sector of the economy. They can and do absorb massive amounts of capital and labour, particularly, as we shall see, under conditions of geographical expansion. Flows into the tertiary circuit of capital—defined as long-term investments in social infrastructures—similarly divide into investment in, say, research and development or skill training that feed directly back into production and those oriented to improving the social condition of the population (through, for example, education and health care). In advanced capitalist countries this last category (e.g. the health–care budget) often absorbs huge amounts of capital. A portion of this investment may also be considered to be in effect geographically immobile. An education system, for example, is hard to move around once it is organized administratively and financially within a given space.

Surpluses generated in the present can be and are absorbed into the secondary and tertiary circuits of capital. These investments can be productive in the long run if they contribute to the future productivity of capital. This occurs if a more educated labour force, investment in research and development, or a more efficient transport and communications system eases the path to further capital accumulation. If this is the case, then overaccumulated capital eventually flows back into the primary circuit of capital, but it may take many years to do so and by then another round of investment in physical and social infrastructures may be called for. Investments of this sort offer relief, at least for a time, for the overaccumulation problem. But over-investment in the secondary and tertiary circuits of capital can also occur, in which case there will be surpluses of housing, office space, and factory and port facilities, as well as excess capacity in, say, the educational system. In this case assets will end up devalued within the secondary or tertiary circuits themselves.

Overaccumulation within the secondary and tertiary circuits often acts as a trigger for more general crises. The importance of this is all too often neglected in general accounts of the dynamics of capital accumulation (Brenner, for one, ignores it). For example, the starting point of the crisis of 1973–5 was a world-wide collapse of property markets followed shortly thereafter by the virtual bankruptcy of New York City; the beginning of the decade-long stagnation in Japan in 1990 was a collapse of the speculative bubble in land, property, and other asset prices, putting the whole banking system in jeopardy (interestingly, the Japanese government sought to compensate periodically by massive state expenditures on public works); the beginning of the Asian collapse in 1997 was the bursting of the property bubbles in Thailand and Indonesia; and the most important prop to the US and British economies after the onset of general recession in all other sectors from mid-2001 onwards was the continued speculative vigour in the property and housing markets and construction. In a curious backwash effect, we find that some 20 per cent of GDP growth in the United States in 2002 was attributable to consumers refinancing their mortgage debt on the inflated values of their housing and using the extra money they gained for immediate consumption (in effect, mopping up overaccumulating capital in the primary circuit). British consumers borrowed $19 billion in the third quarter of 2002 alone against the value of their mortgages to finance their consumption. What happens if and when this property bubble bursts is a matter for serious concern.[18] We have also to consider the possible impact of the vast programme of public works that the Chinese government is currently contemplating as one possible way in which global over-accumulation will find at least a partial outlet in the near future (in much the same way that the interstate highway system and all its ancillary work of suburbanization and the development of the south and west in the United States helped absorb surplus capitals in the 1950s and 1960s).

But all of this depends upon the crucial mediating role of financial and/or state institutions in switching flows of capital between the three circuits of capital. Surplus capital in shirts and shoes cannot be converted directly into an airport or research institute. State and financial institutions have the key power to generate and offer credit. They in effect create a quantity of what may be called 'fictitious capital' (paper assets or promissory notes that have no material backing but which can be used as money).[19] Suppose they create fictitious capital roughly equivalent to the excess capital locked into the production of shirts and shoes and switch it into future-oriented projects in, say, highway construction or education, thereby reinvigorating the economy (including, perhaps, augmenting the demand for shirts and shoes by teachers and construction workers). If the expenditures on built environments or social improvements prove productive (i.e. facilitative of more efficient forms of capital accumulation later on) then the fictitious values are redeemed (either directly by retirement of debt or indirectly in the form of, say, higher tax returns to pay off state debt). The theory of productive state expenditures that pay for themselves out of growth and higher tax yields has frequently been put into practice, as in the case of the remaking of Paris during the Second Empire.[20] But the theory does not always work, and over-investment in built environments or in social expenditures can result in devaluations of these assets or difficulties in paying off state debts. During the 1960s in the United States, for example, it was believed that massive investment in education would pay off in the long run and create a new basis for further accumulation. This broadly failed to happen, and the fiscal crisis of the US state (including that of New York City) that matured during the 1970s was partly due to over-investment in the production of physical and social infrastructures of this kind (the cost of the war in Vietnam being the other part of the problem).

Even in the face of fiscal failure, such investments may prove of inestimable worth in the end because many of them stay in existence as physical use values. Surplus capital largely from the United States (Baltimore in particular) went into the construction of a lot of the London underground system at the beginning of the twentieth century, which promptly went bankrupt but which left the tunnels in place for subsequent generations to use. The classic tale in this regard is the property company Olympia & York, which made its fortune buying up bankrupt properties at fire-sale prices and then turning them into going propositions. Olympia & York came unstuck when it launched its own project at Canary Wharf and was foreclosed upon by the banks given the failure of the project to realize an adequate rate of return. The banks wrote down the value of the property and sold it to investors, who seem to have done very well out of the project ever since (Olympia & York, realizing this possibility, became part of a consortium to buy back the property at the lower price!). As Marx presciently observed, the first wave of investors frequently goes bankrupt in such endeavours, leaving the profitable business to accrue to those who buy up the devalued assets at rock-bottom prices. The devaluation of assets, particularly in the secondary circuit of capital, can, therefore, play an important role in establishing a fresh basis for capital accumulation.

The Spatio-Temporal Fix

The term 'fix' has a double meaning in my argument. A certain portion of the total capital is literally fixed in and on the land in some physical form for a relatively long period of time (depending on its economic and physical lifetime). Some social expenditures (such as public education or a health-care system) also become territorialized and rendered geographically immobile through state commitments. The spatio-temporal 'fix', on the other hand, is a metaphor for a particular kind of solution to capitalist crises through temporal deferral and geographical expansion. So how and when do these material and metaphorical meanings collide?

Capital Bondage

The production of space, the organization of wholly new territorial divisions of labour, the opening up of new and cheaper resource complexes, of new regions as dynamic spaces of capital accumulation, and the penetration of pre-existing social formations by capitalist social relations and institutional arrangements (such as rules of contract and private property arrangements) provide important ways to absorb capital and labour surpluses. Such geographical expansions, reorganizations, and reconstructions often threaten, however, the values already fixed in place (embedded in the land) but not yet realized. This contradiction is inescapable, and open to endless repetition because new regions also require fixed capital in physical infrastructures and built environments if they are to function effectively. The vast quantities of capital fixed in place act as a drag upon the capacity to realize a spatial fix elsewhere. The value of the assets that constitute New York City were and are not trivial and the threat of their devaluation in 1975 (and now again in 2003) was (and is) properly viewed as a major threat not only to the city but to the whole future of capitalism. If capital does move out, then it leaves behind a trail of devastation and devaluation; the deindustrializations experienced in the heartlands of capitalism (such as Pittsburgh, Sheffield, the Ruhr), as well as in many other parts of the world (such as Bombay), in the 1970s and 1980s are cases in point. If capital does not or cannot move, on the other hand, then overaccumulated capital stands to be devalued directly through the onset of a deflationary recession or depression.

Contradictions arise, however, within the dynamics of spatio-temporal transformations. If the surpluses of capital and of labour power exist within a given territory (such as a nation-state or a region) and cannot be absorbed internally (either by geographical adjustments or social expenditures) then they must be sent elsewhere to find a fresh terrain for their profitable realization if they are not to be devalued. This can happen in a number of ways. Markets for commodity surpluses can be found elsewhere. But the spaces to which the surpluses are sent must possess means of payment such as gold or currency (e.g. dollar) reserves or tradeable commodities. Surpluses of commodities are sent out and money or commodities flow back. The problem of overaccumulation is alleviated only in the short term (it merely switches the surplus from commodity to money or into different commodity forms, though if the latter turn out, as is often the case, to be cheaper raw materials or other inputs they can open up new opportunities for profit-making). If the territory does not possess reserves or commodities to trade back, it must either find them (as Britain forced India to do by opening up the opium trade with China in the nineteenth century and thus extracting Chinese silver via Indian-grown opium) or be given credit or aid. In the latter case a foreign territory is lent or donated the money with which to buy the surplus commodities generated at home. The British did this with Argentina in the nineteenth century, and Japanese trade surpluses during the 1990s were largely absorbed by lending to the United States to support the consumerism that purchased Japanese goods (though the US in this case also had the advantage that it prints the dollar as a means of payment and therefore has rights to seigniorage; if it chooses to, it can so regulate the international value of the dollar as to pay the Japanese back in devalued currency). One of the tactics of the US arms industry is to get the government, for reasons of 'security', to lend to a foreign government (most recently Poland) to purchase US-made military equipment. Market and credit transactions of this sort can alleviate problems of overaccumulation within a particular territory, at least in the short term. They function well under conditions of uneven geographical development in which surpluses available in one territory are matched by lack of supply elsewhere.

But resort to the credit system simultaneously makes territories vulnerable to flows of speculative and fictitious capitals that can both stimulate and undermine capitalist development and even, as in recent years, be used to impose savage devaluations upon them. Territorial indebtedness became more and more of a global problem after 1980 or so, and many of the poorer countries (and even some major powers, like Russia in 1998 and Argentina after 2001) found it impossible to pay their debts, threatening default. To deal with this difficulty a permanent organization of nineteen creditor countries, known as the Paris Club, was created to establish rules for debt rescheduling for countries unable to pay off their creditors. Since 2000 some thirty-seven countries have been forced to take this route, and pressure has been growing on the Paris Club to forgive debt entirely for some of the poorest countries. What Cheryl Payer calls 'the debt trap' has to be seen, however, as a process of 'hooking in' even the poorest countries to the system of capital circulation so that they can be available as 'sinks' for surplus capitals for which they are judged liable.[21] It is the receiving country which has to compensate for any devaluation of capital and the creditor country that is protected from devaluation. The resources of the receiving countries can then easily be plundered under the draconian rules of debt repayment.

The export of capital, particularly when accompanied by the export of labour power, works rather differently and typically has longer-term effects. In this case, surpluses of capital and labour are sent elsewhere to set capital accumulation in motion in the new regional space. Surpluses of British capital and labour generated in the nineteenth century found their way to the United States, to the settler colonies such as South Africa, Australia, and Canada, creating new and dynamic centres of accumulation in these territories which generated a demand for goods from Britain. US foreign aid in recent times has almost always been tied to the purchase of US goods and services, thereby functioning as a de facto support for the US economy. Since it may take many years for capitalism to mature in these new territories (if it ever does) to the point where they, too, begin to produce overaccumulations of capital, the originating country can hope to benefit from this process for a not inconsiderable period of time. This is particularly the case when the goods demanded elsewhere are to be embedded as fixed capital in the land. Portfolio investments can support the construction of railroads, highways, ports, dams, and other infrastructures required as a basis for robust capital accumulation in the future. But the rate of return on these long-term investments in the built environment eventually depends upon the evolution of a strong dynamic of accumulation in the receiving country (unless, as often happens, the rate of return on the lent capital is guaranteed by the receiving state). Britain lent to

Capital Bondage

Argentina in this way during the last part of the nineteenth century. The United States, via the Marshall Plan for Europe (Germany in particular) and Japan, clearly saw that its own economic security (leaving aside the military aspect of the Cold War) rested on the active revival of capitalist activity in these spaces.

Contradictions arise, as this last case all too amply illustrates, because new dynamic spaces of capital accumulation will ultimately generate surpluses and will seek ways to absorb them through geographical expansions. Japan and Germany became serious competitors against US capital from the late 1960s onwards, much as the US overwhelmed British capital (and helped pull down the British empire) as the twentieth century dragged on. It is always interesting to note the point at which strong internal development spills over into a search for a spatial fix. In Japan it did so during the 1960s, first through trade, then through the export of capital as direct investment, first to the European Union and the United States, more recently by massive investments (both direct and portfolio) in East and South-East Asia in general and China in particular, and finally through lending abroad (particularly to fund the US current account deficit). South Korea suddenly switched outwards in the 1980s, followed by Taiwan in the late 1980s, both countries exporting not only financial capital but some of the most vicious labour management practices imaginable as subcontractors to multinational capital throughout the world (in Central America and Africa, as well as throughout the rest of East and South–East Asia). Even recently successful adherents to capitalist development have, therefore, quickly found themselves in need of a spatio-temporal fix for their overaccumulating capital. The recent rapidity with which certain territories, such as South Korea, Singapore, and Taiwan, moved from being net receiving to net exporting territories has been quite startling relative to the slower rhythms characteristic of former periods. But by the same token these successful territories have to adjust faster to the blowbacks from their own spatio-temporal fixes. China, absorbing surpluses in the form of foreign direct investments from Japan, Korea, and Taiwan, is rapidly supplanting those countries in many lines of production and export.

The generalized over-capacity that Brenner identifies particularly from 1980 onwards can in this way be disaggregated into a hegemonic economic hub (the triad of the United States, Japan, and Europe) and a cascading and proliferating series of spatio-temporal fixes primarily throughout East and South-East Asia but with additional elements within Latin America (Brazil, Mexico, and Chile in particular), supplemented since the end of the Cold War with a series of rapid thrusts into eastern Europe. While these cascading spatio-temporal fixes may be recorded in terms of relationships between territories, they are in fact material and social relations between regionalities built up through the molecular processes of capital accumulation in space and time. The formal territorial difficulties between Taiwan and mainland China appear totally anachronistic when observed against the growing integration of the industrial regions of Taipei and Shanghai.

There are two possible general outcomes to this process. Under the first, new spatio-temporal fixes open up again and again and surplus capitals are absorbed on an episodic basis. What I call 'switching crises' have the effect of redirecting capital flows from one space to another. The capitalist system remains relatively stable as a whole, even though the parts experience periodic difficulties (such as deindustrialization here or partial devaluations there). The overall effect of such inter-regional volatility is to temporarily reduce the aggregate dangers of overaccumulation and devaluation even though localized distress may from time to time be severe. In one sense, the volatility experienced since 1980 or so seems to have largely been of this type, though it was clearly manipulated, if not directed, by the Wall Street–Treasury–IMF complex to the advantage of finance capital, Wall Street, and the US economy. At each step, of course, the issue arises as to which will be the next space into which capital can profitably flow, and why.

In the current conjuncture an obvious candidate to absorb surplus capital is China, and it is useful to look at this briefly since it not only illustrates the potentialities of a contemporary spatio-temporal fix to the overaccumula-tion problem but it also has relevance to the question of shifting hegemony within the global system. China has, of course, become a major recipient of direct foreign investment. Net foreign direct investment rose from $5 billion in 1991 to around $50 billion in 2002. But the China market is also growing very rapidly, with urban incomes rising at a rate of 11 per cent and rural incomes at a rate of 6 per cent a year in recent times. The internal market is growing, as is the market for foreign goods. Not a few multinationals, such as General Motors, made most of their profit out of China sales in 2001–2. The huge potentiality of the internal market in China is not, there-fore, to be ignored and some of the foreign direct invest-

ment in, say, microelectronics is as much oriented to selling internally as it is to exporting to the rest of the world.
But even more dramatic are the prospects for long-term
infrastructural investment. Since 1998, the Chinese have
sought to absorb their vast labour surpluses (and to curb
the threat of social unrest) by debt-financed investment in
huge mega-projects that dwarf the already huge Three
Gorges dam. They are proposing a far more ambitious
project (costing at least $60 billion) to divert water from
the Yangtze to the Yellow River. New subway systems and
highways are being built in major cities, and 8,500 miles of
new railroads are proposed to integrate the interior to the
economically dynamic coastal zone, including a high-
speed link between Shanghai and Beijing and a link into
Tibet. Urban infrastructures are everywhere being
upgraded. The Olympic Games is prompting heavy
investment in Beijing. This effort is far larger *in toto* than
that which the United States undertook during the 1950s
and 1960s, and has the potential to absorb surpluses of
capital for several years to come. It is, however, deficit-
financed, and that entails high risks since if the investments do not return their value to the accumulation
process in due course, then a fiscal crisis of the state will
quickly engulf China with serious consequences for economic development and social stability.[22] Nevertheless,
this proposes to be a remarkable version of a spatio-
temporal fix that has global implications not only for
absorbing overaccumulated capital, but also for shifting
the balance of economic and political power to China as
the regional hegemon and perhaps placing the Asian
region, under Chinese leadership, in a much more competitive position vis-à-vis the United States. All the more

reason, therefore, for the United States to get a handle on the oil supplies that China increasingly needs from the Caspian Basin and from the Middle East.

A second possible outcome, however, is increasingly fierce international competition as multiple dynamic centres of capital accumulation compete on the world stage in the face of strong currents of overaccumulation. Since they cannot all succeed in the long run, either the weakest succumb and fall into serious crises of localized devaluation or geopolitical struggles arise between regions. The latter can get converted via the territorial logic of power into confrontations between states in the form of trade wars and currency wars, with the ever-present danger of military confrontations (of the sort that gave us two world wars between capitalist powers in the twentieth century) lurking in the background. In this case, the spatio-temporal fix takes on a much more sinister form as it transmutes into the export of localized and regional devaluations and destruction of capital (of the sort that occurred on a massive scale in East and South-East Asia and in Russia in 1997–8). How and when this occurs depends, however, just as much upon the explicit forms of political action on the part of state powers as it does upon the molecular processes of capital accumulation in space and time. The dialectic between the territorial logic and the capitalistic logic is now fully engaged. There are, however, some further points to make about this process in order to better understand how it actually works.

Inner Contradictions

In *The Philosophy of Right* Hegel notes how the inner contradictions of bourgeois society, registered as an over-accumulation of wealth at one pole and the creation of a rabble of paupers at the other, drive it to seek solutions through external trade and colonial/imperial practices.[23] In so doing he rejects the idea that there might be ways to solve the problem of social inequality and instability through internal mechanisms of redistribution. Lenin quotes Cecil Rhodes as saying that colonialism and imperialism abroad was the only possible way to avoid civil war at home.[24] Class relations and the state of class struggle within a territorially bounded social formation clearly affect the impetus for a spatio-temporal fix.

The evidence from the end of the nineteenth century is here of interest. Consider, for example, a figure like Joseph Chamberlain ('Radical Joe' as he was known). Closely allied with the liberal manufacturing interests of Birmingham, Chamberlain was initially resolutely opposed to imperialism (in the Afghan Wars of the 1850s, for example) and devoted much of his time to educational reform and other projects aligned to improving the social and physical infrastructures for production and consumption in his home city of Birmingham. This provided, he thought, a productive outlet for surpluses that would be repaid in the long run. An important figure within the liberal conservative movement, he saw the rising tide of class struggle in Britain at first hand, and in 1885 made a celebrated speech in which he called for the propertied classes to take cognizance of their responsibilities and obligations to society (i.e. to better the conditions of life of the least well off and invest in social and physical infra-structures in the national interest) rather than solely to promote their individual rights as property owners. The uproar that followed on the part of the propertied classes forced him to recant, and from that moment on he became the most ardent advocate for imperialism (ultimately as Colonial Secretary, leading Britain into the disaster of the Boer War in South Africa). This sort of career trajectory was quite common for the period. Jules Ferry in France, an ardent supporter of internal reform (particularly education) in the 1860s, took to colonial advocacy after the Commune of 1871 (leading France into the mire of South-East Asia that culminated in defeat at Dien Bien-Phu in 1954), and even Theodore Roosevelt in the United States turned, after the famous declaration of Frederic Jackson Turner that the American frontier was now closed (even though it was far from closed to new investment possibilities in the south and west), to sup-porting imperial practices rather than internal reforms.[25]

In all of these cases, the turn to a liberal form of im-perialism (and one that had attached to it an ideology of progress and of a civilizing mission) resulted not from absolute economic imperatives but from the political unwillingness of the bourgeoisie to give up on any of its privileges and thereby absorb overaccumulation inter-nally through social reform at home, even in the face of growing claims from working-class movements. Hobson, for one, identified this as the key problem and sought a social democratic policy that would counter it.[26] It is, therefore, of critical importance to consider the internal role of class relations and of class struggle, and the particular pattern of class alliances that is constructed within the state (including a class alliance of workers and capitalists around imperial endeavours), in assessing the impetus for imperialist endeavours and the drive outwards to find spatio-temporal fixes. It was internal politics of this sort that forced many European powers to look outwards to solve their problems from 1884 to 1945, and this gave a specific coloration to the forms that European imperialism took during these years. It is surprising to note, for example, how many liberal and even radical figures became proud imperialists and how much of the working-class movement collaborated with the imperial project. This required, however, that bourgeois interests should thoroughly command state policy and military power. I therefore think Arendt is correct, as I argued in Chapter 2, to interpret the imperialism that emerged at the end of the nineteenth century as 'the first stage in political rule of the bourgeoisie rather than the last stage of capitalism' as Lenin depicted it.[27] This is, however, a matter to which we will return in the Chapter 5.

The Powers of Mediating Institutions

The critical mediating role of financial and institutional arrangements and powers (particularly those of the state) in processes of capital accumulation is important to acknowledge. This requires, however, careful scrutiny of the different forms that such mediating institutions might assume and the consequent effects upon the molecular processes of capital accumulation in space and time. In his study of how the crisis of 1997–8 unfolded in East and South-East Asia, for example, Henderson shows that the difference between Taiwan and Singapore (which escaped relatively unscathed except for currency devaluation) and Thailand and Indonesia (which suffered almost total economic and political collapse), turned on differences in state and financial policies.[28] The former countries were insulated from speculative flows by strong states and protected financial markets, whereas the latter, which had liberalized their capital markets, were not. Differences of this sort plainly matter a great deal. In this case, they effectively determined who got hit by savage devaluation and who did not.

On this point, I cannot do much more here than acknowledge the political importance of this issue. Clearly, the whole pattern of turbulence in the relations between state, supra-state, and financial powers on the one hand and the more general dynamics of capital accumulation (through production and selective devaluations) on the other has proven one of the most signal, and most complex, elements in the narrative of uneven geographical development and imperialist politics to be told of the period since 1973. I think Gowan is correct to see the radical restructuring of international capitalism after that date as a series of desperate gambles on the part of the United States to maintain its hegemonic position in world economic affairs against Europe, Japan, and later East and South-East Asia more generally This began during the crisis of 1973 with Nixon's double strategy of high oil pricing and financial deregulation. The US banks were then given the exclusive right to recycle the vast quantities of petrodollars being accumulated in the Gulf region.[29] This recentred global financial activity in the US and incidentally helped, when coupled with the internal reforms of the financial system within the United States, to rescue New York from its own local economic crisis. This resulted in the emergence of a powerful Wall Street/US Treasury financial regime with controlling powers over global financial institutions (such as the IMF) and an ability to make or break many weaker foreign economies through credit manipulations and debt management practices. This monetary and financial regime was used, Gowan goes on to argue, by successive US administrations 'as a formidable instrument of economic statecraft to drive forward both the globalization process and the associated neo-liberal domestic transformations'. The regime thrived on crises: 'The IMF covers the risks and ensures that the US banks don't lose (countries pay up through structural adjustments etc.) and flight of capital from a localized crises elsewhere ends up boosting the strength of Wall Street . . .'.[30] The effect was to project US financial power outwards (in alliance with others wherever possible), to force open markets, particularly for capital and financial flows (now a US-imposed requirement for state membership in the IMF system), and impose other neo-liberal practices (culminating in the WTO) upon much of the rest of the world.

There are two major points to be made about this system. First, free trade in commodities is often depicted as opening up the world to free and open competition. But we have already seen that it necessarily gives rise, when grounded in space, to monopolistic competition, generating asymmetries in exchange even under the best of conditions. The whole argument fails, as Lenin long ago pointed out, in the face of concentrated monopoly or oligopoly power (either in production or consumption). The

US, for example, has repeatedly used the weapon of denial of access to its huge market to force other nations to comply with its wishes. This is a gargantuan version of the asymmetry in exchange that always attaches to space relations. The most recent (and crass) example of this line of argument comes from the US trade representative Robert Zoellick to the effect that, if Lula, the newly elected Workers' Party president of Brazil, does not go along with US plans for free markets in the Americas, he would find himself having 'to export to Antarctica'.[31] Taiwan and Singapore were forced (as Korea was earlier as part of the IMF bailout at the behest of the US Treasury), against their better judgement, to open their financial markets to speculative capital, even though they had earlier been protected from devaluation by keeping their markets closed. They were forced to sign on to the WTO in the face of US threats to deny them access to its market. The US now plans to attach a condition of open market access on the US model to the 'Millennium Challenge Grants' of foreign aid it offers to poor countries. In return for aid, these countries must adopt institutional arrangements compatible with those of the US and thereby lay themselves open to whatever the superior powers of monopolized capital wish or need to do. On the production side, oligopolies largely based in the core capitalists regions effectively control the production of seeds, fertilizers, electronics, computer software, pharmaceutical products, petroleum products, and much more. Under these conditions, the creation of new market openings does not open up competition but merely creates opportunities for monopoly powers to proliferate, with all manner of social, ecological, economic, and political consequences. This is as true for the export of multinational capitals to produce shoes and shirts throughout South-East Asia and Latin America as it is for the marketing of Coca-Cola. Even something as seemingly benevolent as the Green Revolution has, most commentators agree, paralleled the increased agricultural outputs with considerable concentrations of wealth in the agrarian sector and higher levels of dependency upon monopolized inputs throughout East and South–East Asia. The penetration of the China market by US tobacco companies is set fair to compensate their losses in the US market at the same time as it will surely generate a public health crisis in China for decades to come. In all of these respects, the claims generally made for neo–liberalism to be about open rather than monopolistic competition, to be about fair as well as free trade, turn out to be fraudulent, masked as usual by the fetishism of the market.

There is also, as even advocates of free trade readily acknowledge, a huge difference between freedom of trade in commodities and freedom of movement for finance capital. This immediately poses the problem of what kind of market freedom is being talked about. Some, like Bhagwati, fiercely defend free trade in commodities but resist the idea that this necessarily holds good for financial flows.[32] The difficulty here is this. On the one hand credit flows are vital to productive investments and reallocations of capital from one line of production or location to another. They also play an important role in bringing consumption needs (for housing, for example) into a potentially balanced relationship with productive activities in a spatially disaggregated world marked by surpluses in one space and deficits in another. In all of these respects the financial system (with or without state involvement) is critical to coordinate the dynamics of capital accumulation. But finance capital also embraces a lot of unproductive activity in which money is simply used to make more money through speculation on commodity futures, currency values, debts, and the like. When huge quantities of capital become available for such purposes, then open capital markets become vehicles for speculative activity, some of which, as we saw during the 1990s with both the dot.com and the stock market 'bubbles', become self-fulfilling prophecies, just as the hedge funds, armed with trillions of dollars of leveraged money, could force Indonesia and even Korea into bankruptcy no matter what the strength of their underlying economies. Much of what happens on Wall Street has nothing to do with facilitating investment in productive activities. It is purely speculative (hence the descriptions of it as 'casino' or even 'vulture' capitalism). But this activity has deep impacts upon the overall dynamics of capital accumulation, and most particularly on the recentring of political–economic power primarily in the United States but also within the financial markets of other core countries (Tokyo, London, Frankfurt).

The State Steps Back In

It is at this point that the territorialized politics of state and empire re-enter to claim a leading role in the continuing drama of endless capital accumulation and overaccumulation. It is the state that is the political entity, the body politic, that is best able to orchestrate institutional arrangements and manipulate the molecular forces of

capital accumulation to preserve that pattern of asymmetries in exchange that are most advantageous to the
dominant capitalist interests working within its frame. If,
for example, we find that the WTO proclaims free trade
but actually delivers unfair trade in which the richer
countries maintain their collective advantage over the
poorer, then we should not be surprised. This is typical of
imperial practices. Britain insisted upon free (and unfair)
trade and laissez-faire during the nineteenth century
when it was to its advantage so to do, but abandoned such
a posture as soon as the benefits began to accrue to others.
The United States subsequently took up the banner of
first the 'open door' but then free trade to the point where
the current rhetoric of the Bush administration equates
freedom with free trade without a hint of any possible
incompatibility between freedoms of self-determination
on the one hand and the imposed discipline of free
markets and unfair trade on the other. Imperialism, in this
domain, amounts to foisting institutional arrangements
and conditions upon others, usually in the name of universal well-being. This is the central thrust of the Bush
administration's current policies, as I noted in Chapter 1.
'We seek,' says President Bush as he goes to war, 'a just
peace where repression, resentment and poverty are
replaced with the hope of democracy, development, free
markets and free trade.' These last two have 'proved their
ability to lift whole societies out of poverty'. The United
States will deliver this gift of freedom (of the market) to
the world whether it likes it or not.

How all this actually occurs depends critically on the
nature of governance and the dominant form of the class
alliances, particularly within the core countries which

initially produce and then control the disbursement of surplus capitals. These countries have a disproportionate influence upon the financial architecture through which spatio-temporal fixes are predominantly pursued, and are therefore in a position to calibrate the inevitable asymmetries that exist in spatial exchange to their own advantage. The emergence of a 'Wall Street–Treasury' complex within the United States, able to control institutions such as the IMF and to project vast financial power across the world through a network of other financial and governmental institutions, has exercised massive influence over the dynamics of global capitalism in recent years. But this power centre can only operate in the way it does because the rest of the world is networked and successfully hooked into (and effectively 'hooked on' usually by way of credit arrangements) a structured framework of interlocking financial and governmental (including supra–national) institutions.

The general picture which then emerges, is of a networked spatio-temporal world of financial flows of surplus capital with conglomerations of political and economic power at key nodal points (New York, London, Tokyo) seeking either to disburse and absorb the surpluses down productive paths, more often than not in long-term projects across a variety of spaces (from Bangladesh to Brazil or China), or to use speculative power to rid the system of overaccumulation by the visitation of crises of devaluation upon vulnerable territories. It is, of course, the populations of those vulnerable territories who then must pay the inevitable price, in terms of loss of assets, loss of jobs, and loss of economic security, to say nothing of the loss of dignity and hope. And by the same logic that has it that

the most vulnerable territories get hit first, so it is typically the most vulnerable populations within those territories that bear the brunt of any burden. It was the rural poor
of Mexico, Thailand, and Brazil who suffered most from
the depredations that flowed from the financial crises of
the 1980s and 1990s. Capitalism survives, therefore, not
only through a series of spatio-temporal fixes that absorb
the capital surpluses in productive and constructive ways,
but also through the devaluation and destruction administered as corrective medicine to what is generally
depicted as the fiscal profligacy of those who borrow. The
very idea that those who irresponsibly lend might also be
held responsible is, of course, dismissed out of hand
by ruling elites. That would require calling the wealthy
property-owning classes everywhere to account and
insisting that they look to their responsibilities rather than
to their inalienable rights to private property and a satis-
factory rate of profit. But, as Joseph Chamberlain found,
it is far easier politically to pillage and debase far-away
populations (particularly those who are racially, ethnically, or culturally different), than to confront overwhelming capitalist class power at home. The sinister and
destructive side of spatial-temporal fixes to the overaccumulation problem becomes just as crucial an element
within the historical geography of capitalism as does its
creative counterpart in building a new landscape to
accommodate both the endless accumulation of capital
and the endless accumulation of political power.

 If the official rhetoric is to be believed, the complex of
institutional arrangements that now mediate flows of
capital around the world should be geared to sustain and
support expanded reproduction (growth), to ward off any

trend towards crises, and to seriously address the problem of poverty reduction. But, if that project fails, it can seek to accumulate by other means. Like war in relation to diplomacy, finance capital intervention backed by state power frequently amounts to accumulation by other means. An unholy alliance between state powers and the predatory aspects of finance capital forms the cutting edge of a 'vulture capitalism' that is as much about cannibalistic practices and forced devaluations as it is about achieving harmonious global development. But how are we to interpret these 'other means' to accumulation?

4

Accumulation by Dispossession

Rosa Luxemburg argues that capital accumulation has a dual character:

One concerns the commodity market and the place where surplus value is produced—the factory, the mine, the agricultural estate. Regarded in this light accumulation is a purely economic process, with its most important phase a transaction between the capitalist and the wage labourer. . . . Here, in form at any rate, peace, property and equality prevail, and the keen dialectics of scientific analysis were required to reveal how the right of ownership changes in the course of accumulation into appropriation of other people's property, how commodity exchange turns into exploitation, and equality becomes class rule. The other aspect of the accumulation of capital concerns the relations between capitalism and the non-capitalist modes of production which start making their appearance on the international stage. Its predominant methods are colonial policy, an international loan system—a policy of spheres of interest—and war. Force, fraud, oppression, looting are openly displayed without any attempt at concealment, and it requires an effort to discover within this tangle of political violence and contests of power the stern laws of the economic process.[1]

137

These two aspects of accumulation, she argues, are 'organically linked' and 'the historical career of capitalism can only be appreciated by taking them together'.

Underconsumption or Overaccumulation?

Luxemburg rests her analysis upon a particular understanding of the crisis tendencies of capitalism. The problem, she argues, is underconsumption, a general lack of sufficient effective demand to soak up the growth in output that capitalism generates. This difficulty arises because workers are exploited and by definition receive much less value to spend than they produce, and capitalists are at least in part obliged to reinvest rather than to consume. After due consideration of various ways in which the supposed gap between supply and effective demand might be bridged, she concludes that trade with non-capitalist social formations provides the only systematic way to stabilize the system. If those social formations or territories are reluctant to trade then they must be compelled to do so by force of arms (as happened with the opium wars in China). This is, in her view, the heart of what imperialism is about. One possible corollary of this argument (though Luxemburg does not state it directly) is that, if this system is to last any length of time, the non-capitalist territories must be kept (forcibly if necessary) in a non–capitalist state. This could account for the fiercely repressive qualities of many of the colonial regimes developed during the latter half of the nineteenth century.

Few would now accept Luxemburg's theory of underconsumption as the explanation of crises.[2] By contrast, the theory of overaccumulation identifies the lack of opportunities for profitable investment as the fundamental problem. On occasion, lack of sufficient effective consumer demand may be part of the problem—hence the heavy reliance in our own day on something called 'consumer confidence' (otherwise known as the inability of compulsive shoppers to keep their credit cards in their wallets) as an indicator of strength and stability in the economy. The gap that Luxemburg thought she saw can easily be covered by reinvestment which generates its own demand for capital goods and other inputs. And, as we have seen in the case of the spatio-temporal fixes, the geographical expansion of capitalism which underlies a lot of imperialist activity is very helpful to the stabilization of the system precisely because it opens up demand for both investment goods and consumer goods elsewhere. Imbalances can arise, of course, between sectors and regions, and business cycles and localized recessions can result. But it is also possible to accumulate in the face of stagnant effective demand if the costs of inputs (land, raw materials, intermediate inputs, labour power) decline significantly. Access to cheaper inputs is, therefore, just as important as access to widening markets in keeping profitable opportunities open. The implication is that non-capitalist territories should be forced open not only to trade (which could be helpful) but also to permit capital to invest in profitable ventures using cheaper labour power, raw materials, low-cost land, and the like. The general thrust of any capitalistic logic of power is not that territories should be held back from capitalist development, but that they should be continuously opened up. From this standpoint colonial repressions of the sort that undoubtedly occurred in the late nineteenth century have to be interpreted as self-defeating, a case of a territorial logic inhibiting the capitalistic logic. Fear of emulation led Britain, for example, to prevent India from developing a vigorous capitalist dynamic and thereby frustrated the possibilities of spatio-temporal fixes in that region. The open dynamic of the Atlantic economy did far more for Britain than did the repressed colonial empire in India, from which Britain certainly managed to extract surpluses but which never functioned as a major field for deployment of British surplus capital. But, by the same token, it was the open dynamic of the Atlantic trade that opened up the possibility of Britain's displacement by the United States as the global hegemonic power. If Arendt is right and endless accumulation requires the endless accumulation of political power, then such shifts are impossible to avoid and any attempt to do so will result in disaster. The formation of closed empires after the First World War almost certainly played a role in the inability to solve the overaccumulation problem of the 1930s and laid the economic groundwork for the territorial conflicts of the Second World War. The territorial logic dominated and frustrated the capitalist logic, thus forcing the latter into an almost terminal crisis through territorial conflict.

The weight of historical-geographical evidence from the twentieth century broadly accords with the overaccumulation argument. However, there is much that is interesting about Luxemburg's formulation. To begin with, the idea that capitalism must perpetually have something 'outside of itself' in order to stabilize itself is worthy of scrutiny, particularly as it echoes Hegel's conception, which we encountered in Chapter 3, of an inner dialectic of capitalism forcing it to seek solutions external to itself. Consider, for example, Marx's argument concerning the creation of an industrial reserve army.[3] Capital accumulation, in the absence of strong currents of labour-saving technological change, requires an increase in the labour force. This can come about in a number of ways. Increase of population is important (and most analysts conveniently forget Marx's own strictures on this point). Capital can also raid 'latent reserves' from a peasantry or, by extension, mobilize cheap labour from colonies and other external settings. Failing this, capitalism can utilize its powers of technological change and investment to induce unemployment (lay-offs) thus creating an industrial reserve army of unemployed workers directly. This unemployment tends to exert a downward pressure on wage rates and thereby opens up new opportunities for profitable deployment of capital. Now in all of these instances capitalism does indeed require something 'outside of itself' in order to accumulate, but in the last case it actually throws workers out of the system at one point in time in order to have them to hand for purposes of accumulation at a later point in time. Put in the language of contemporary postmodern political theory, we might say that capitalism necessarily and always creates its own 'other'. The idea that some sort of 'outside' is necessary for the stabilization of capitalism therefore has relevance. But capitalism can either make use of some pre-existing outside (non-capitalist social formations or some sector within capitalism—such as education—that has not yet been proletarianized) or it can actively manufacture it. I propose to take this 'inside–outside' dialectic seriously in what follows. I shall examine how the 'organic relation' between expanded reproduction on the one hand and the often violent processes of dispossession on the other have shaped the historical geography of capitalism. This helps us better understand what the capitalistic form of imperialism is about.

Arendt, interestingly, advances an argument along similar lines. The depressions of the 1860s and 1870s in Britain, she argues, initiated the push into a new form of imperialism:

Imperialist expansion had been touched off by a curious kind of economic crisis, the overproduction of capital and the emergence of 'superfluous' money, the result of oversaving, which could no longer find productive investment within the national borders. For the first time, investment of power did not pave the way for investment of money, but export of power followed meekly in the train of exported money, since uncontrolled investments in distant countries threatened to transform large strata of society into gamblers, to change the whole capitalist economy from a system of production into a system of financial speculation, and to replace the profits of production with profits in commissions. The decade immediately before the imperialist era, the seventies of the last century, witnessed an unparalleled increase in swindles, financial scandals and gambling in the stock market.

This scenario sounds all too familiar given the experience of the 1980s and 1990s. But Arendt's description of the bourgeois response is even more arresting. They realized, she argues, 'for the first time that the original sin of simple robbery, which centuries ago had made possible "the original accumulation of capital" (Marx) and had started all further accumulation, had eventually to be repeated lest the motor of accumulation suddenly die down'.[4]

Accumulation by Dispossession

The processes that Marx, following Adam Smith, referred to as 'primitive' or 'original' accumulation constitute, in Arendt's view, an important and continuing force in the historical geography of capital accumulation through imperialism. As in the case of labour supply, capitalism always requires a fund of assets outside of itself if it is to confront and circumvent pressures of overaccumulation. If those assets, such as empty land or new raw material sources, do not lie to hand, then capitalism must somehow produce them. Marx, however, does not consider this possibility except in the case of the creation of an industrial reserve army through technologically induced unemployment. It is interesting to consider why.

Marx's Reticence

Marx's general theory of capital accumulation is constructed under certain crucial initial assumptions that broadly match those of classical political economy. These assumptions are: freely functioning competitive markets with institutional arrangements of private property, juridical individualism, freedom of contract, and appropriate structures of law and governance guaranteed by a 'facilitative' state which also secures the integrity of money as a store of value and as a medium of circulation. The role of the capitalist as a commodity producer and exchanger is already well established, and labour power has become a commodity that trades generally at its appropriate value. 'Primitive' or 'original' accumulation has already occurred and accumulation now proceeds as expanded reproduction (albeit through the exploitation of living labour in production) under conditions of 'peace, property and equality'. These assumptions allow us to see what will happen if the liberal project of the classical political economists or, in our times, the neo-liberal project of the economists, is realized. The brilliance of Marx's dialectical method, as Luxemburg for one clearly recognizes, is to show that market liberalization—the credo of the liberals and the neo-liberals—will not produce a harmonious state in which everyone is better off. It will instead produce ever greater levels of social inequality (as indeed has been the global trend over the last thirty years of neo-liberalism, particularly within those countries such as Britain and the United States that have most closely hewed to such a political line). It will also, Marx predicts, produce serious and growing instabilities culminating in chronic crises of overaccumulation (of the sort we are now witnessing).

The disadvantage of these assumptions is that they relegate accumulation based upon predation, fraud, and violence to an 'original stage' that is considered no longer relevant or, as with Luxemburg, as being somehow 'outside of' capitalism as a closed system. A general re-evaluation of the continuous role and persistence of the predatory practices of 'primitive' or 'original' accumulation within the long historical geography of capital accumulation is, therefore, very much in order, as several commentators have recently observed.[5] Since it seems peculiar to call an ongoing process 'primitive' or 'original' I shall, in what follows, substitute these terms by the concept of 'accumulation by dispossession'.

Accumulation by Dispossession

A closer look at Marx's description of primitive accumulation reveals a wide range of processes.[6] These include the commodification and privatization of land and the forceful expulsion of peasant populations; the conversion of various forms of property rights (common, collective, state, etc.) into exclusive private property rights; the suppression of rights to the commons; the commodification of labour power and the suppression of alternative (indigenous) forms of production and consumption; colonial, neo-colonial, and imperial processes of appropriation of assets (including natural resources); the monetization of exchange and taxation, particularly of land; the slave trade; and usury, the national debt, and ultimately the credit system as radical means of primitive accumulation. The state, with its monopoly of violence and definitions of legality, plays a crucial role in both backing and promoting these processes and, as I argued in Chapter 3, there is considerable evidence that the transition to capitalist development was and continues to be vitally contingent upon the stance of the state. The developmental role of the state goes back a long way, keeping the territorial and capitalistic logics of power always intertwined though not necessarily concordant.

All the features of primitive accumulation that Marx mentions have remained powerfully present within capitalism's historical geography up until now. Displacement of peasant populations and the formation of a landless proletariat has accelerated in countries such as Mexico and India in the last three decades, many formerly common property resources, such as water, have been privatized

145

(often at World Bank insistence) and brought within the capitalist logic of accumulation, alternative (indigenous and even, in the case of the United States, petty commodity) forms of production and consumption have been suppressed. Nationalized industries have been privatized. Family farming has been taken over by agribusiness. And slavery has not disappeared (particularly in the sex trade).

Critical engagement over the years with Marx's account of primitive accumulation—which in any case had the quality of a sketch rather than a systematic exploration—suggests some lacunae that need to be remedied. The process of proletarianization, for example, entails a mix of coercions and of appropriations of pre-capitalist skills, social relations, knowledges, habits of mind, and beliefs on the part of those being proletarianized. Kinship structures, familial and household arrangements, gender and authority relations (including those exercised through religion and its institutions) all have their part to play. In some instances the pre-existing structures have to be violently repressed as inconsistent with labour under capitalism, but multiple accounts now exist to suggest that they are just as likely to be co-opted in an attempt to forge some consensual as opposed to coercive basis for working-class formation. Primitive accumulation, in short, entails appropriation and co-optation of pre-existing cultural and social achievements as well as confrontation and supersession. The conditions of struggle and of working-class formation vary widely and there is, therefore, as Thompson among others has insisted, a sense in which a working class 'makes itself' though never, of course, under conditions of its own choosing.[7] The result is often to leave a trace of pre-capitalist social relations in working–class formation and to create distinctive geographical, historical, and anthropological differentiations in how a working class is defined. No matter how universal the process of proletarianization, the result is not the creation of a homogeneous proletariat.[8]

Some of the mechanisms of primitive accumulation that Marx emphasized have been fine-tuned to play an even stronger role now than in the past. The credit system and finance capital became, as Lenin, Hilferding, and Luxemburg all remarked at the beginning of the twentieth century, major levers of predation, fraud, and thievery. The strong wave of financialization that set in after 1973 has been every bit as spectacular for its speculative and predatory style. Stock promotions, ponzi schemes, structured asset destruction through inflation, asset-stripping through mergers and acquisitions, and the promotion of levels of debt incumbency that reduce whole populations, even in the advanced capitalist countries, to debt peonage, to say nothing of corporate fraud and dispossession of assets (the raiding of pension funds and their decimation by stock and corporate collapses) by credit and stock manipulations—all of these are central features of what contemporary capitalism is about. The collapse of Enron dispossessed many of their livelihoods and their pension rights. But above all we have to look at the speculative raiding carried out by hedge funds and other major institutions of finance capital as the cutting edge of accumulation by dispossession in recent times.

Wholly new mechanisms of accumulation by dispossession have also opened up. The emphasis upon intellectual property rights in the WTO negotiations (the so-called TRIPS agreement) points to ways in which the patenting and licensing of genetic material, seed plasma, and all manner of other products can now be used against whole populations whose practices had played a crucial role in the development of those materials. Biopiracy is rampant and the pillaging of the world's stockpile of genetic resources is well under way to the benefit of a few large pharmaceutical companies. The escalating depletion of the global environmental commons (land, air, water) and proliferating habitat degradations that preclude anything but capital-intensive modes of agricultural production have likewise resulted from the wholesale commodification of nature in all its forms. The commodification of cultural forms, histories, and intellectual creativity entails wholesale dispossessions (the music industry is notorious for the appropriation and exploitation of grassroots culture and creativity). The corporatization and privatization of hitherto public assets (such as universities), to say nothing of the wave of privatization (of water and public utilities of all kinds) that has swept the world, indicate a new wave of 'enclosing the commons'. As in the past, the power of the state is frequently used to force such processes through even against popular will. The rolling back of regulatory frameworks designed to protect labour and the environment from degradation has entailed the loss of rights. The reversion of common property rights won through years of hard class struggle (the right to a state pension, to welfare, to national health care) to the private domain has been one of the most egregious of all policies of dispossession pursued in the name of neo-liberal orthodoxy.

Capitalism internalizes cannibalistic as well as predatory and fraudulent practices. But it is, as Luxemburg cogently observed, 'often hard to determine, within the tangle of violence and contests of power, the stern laws of the economic process'. Accumulation by dispossession can occur in a variety of ways and there is much that is both contingent and haphazard about its *modus operandi*.

So how, then, does accumulation by dispossession help solve the overaccumulation problem? Overaccumulation, recall, is a condition where surpluses of capital (perhaps accompanied by surpluses of labour) lie idle with no profitable outlets in sight. The operative term here, however, is the capital surplus. What accumulation by dispossession does is to release a set of assets (including labour power) at very low (and in some instances zero) cost. Overaccumulated capital can seize hold of such assets and immediately turn them to profitable use. In the case of primitive accumulation as Marx described it, this entailed taking land, say, enclosing it, and expelling a resident population to create a landless proletariat, and then releasing the land into the privatized mainstream of capital accumulation. Privatization (of social housing, telecommunications, transportation, water, etc. in Britain, for example) has, in recent years, opened up vast fields for overaccumulated capital to seize upon. The collapse of the Soviet Union and then the opening up of China entailed a massive release of hitherto unavailable assets into the mainstream of capital accumulation. What would have happened to overaccumulated capital these last thirty years if these new terrains of accumulation had not opened up? Put another way, if capitalism has been experiencing a chronic difficulty of overaccumulation since 1973, then the neo-liberal project of privatization of everything makes a lot of sense as one way to solve the problem. Another way would be to release cheap raw materials (such as oil) into the system. Input costs would be reduced and profits thereby enhanced. As the news-paper baron Rupert Murdoch observed, the solution to our current economic woes is oil at $20 rather than $30 or more a barrel. Small wonder that all of Murdoch's news-papers have been such avid supporters of war against Iraq.[9]

The same goal can be achieved, however, by the devalu-ation of existing capital assets and labour power. Devalued capital assets can be bought up at fire-sale prices and profitably recycled back into the circulation of capital by overaccumulated capital. But this requires a prior wave of devaluation, which means a crisis of some kind. Crises may be orchestrated, managed, and controlled to rational-ize the system. This is often what state-administered austerity programmes, making use of the key levers of interest rates and the credit system, are often all about. Limited crises may be imposed by external force upon one sector or upon a territory or whole territorial complex of capitalist activity. This is what the international financial system (led by the IMF) backed by superior state power (such as that of the United States) is so expert at doing. The result is the periodic creation of a stock of devalued, and in many instances undervalued, assets in some part of the world, which can be put to profitable use by the capital surpluses that lack opportunities elsewhere. Wade and Veneroso capture the essence of this when they write of the Asian crisis of 1997–8:

Financial crises have always caused transfers of ownership and power to those who keep their own assets intact and who are in a position to create credit, and the Asian crisis is no exception

. . . there is no doubt that Western and Japanese corporations are the big winners. . . . The combination of massive devaluations, IMF-pushed financial liberalization, and IMF-facilitated recovery may even precipitate the biggest peacetime transfer of assets from domestic to foreign owners in the past fifty years anywhere in the world, dwarfing the transfers from domestic to US owners in Latin America in the 1980s or in Mexico after 1994. One recalls the statement attributed to Andrew Mellon: 'In a depression assets return to their rightful owners.'[10]

Regional crises and highly localized place-based devaluations emerge as a primary means by which capitalism perpetually creates its own 'other' in order to feed upon it. The financial crises of East and South-East Asia in 1997–8 were a classic case of this.[11] The analogy with the creation of an industrial reserve army by throwing people out of work is exact. Valuable assets are thrown out of circulation and devalued. They lie fallow and dormant until surplus capital seizes upon them to breath new life into capital accumulation. The danger, however, is that such crises might spin out of control and become generalized, or that the 'othering' will provoke a revolt against the system that creates it. One of the prime functions of state interventions and of international institutions is to orchestrate devaluations in ways that permit accumulation by dispossession to occur without sparking a general collapse. This is the essence of what a structural adjustment programme administered by the IMF is all about. For the main capitalist powers, such as the United States, this means orchestrating these processes to their specific advantage, while proclaiming their role as that of a noble leader organizing 'bail-outs' (as in Mexico in 1994) to keep global capital accumulation on track. But there is, as with any speculative gamble, a danger of losing: the sudden evident panic of the US Treasury and the IMF in December 1998 after Russia, with nothing left to lose, had simply declared bankruptcy and when it seemed that the South Korean economy (after several months of hard bargaining) was about to crash and possibly spark a global chain reaction, illustrates how close to the edge such forms of calculation can go.[12]

The mixture of coercion and consent within such bargaining activity varies considerably, but we can now more clearly see how hegemony gets constructed through financial mechanisms in such a way as to benefit the hegemon while leading the subaltern states on the supposedly golden path of capitalist development. The umbilical cord that ties together accumulation by dispossession and expanded reproduction is that given by finance capital and the institutions of credit, backed, as ever, by state powers.

The Contingency of It All

How, then, can we uncover the iron laws within the contingencies of accumulation by dispossession? We know, of course, that a certain level of this goes on all the time and that it can take many forms, both legal and illegal. Consider, for example, a process in US housing markets known as 'flipping'. A house in poor condition is bought for next to nothing, given some cosmetic improvements, then sold on at an exorbitant price, with the aid of a mortgage package arranged by the seller, to a low-income family looking to realize its dream of home ownership. If the family has difficulty meeting the payments or dealing with the serious maintenance problems that almost certainly emerge, then the house is repossessed. This is not exactly illegal (buyers beware!) but the effect is to prey upon low-income families and bilk them of whatever little savings they have. This is accumulation by dispossession. There are innumerable activities (legal and illegal) of this kind that affect the control of assets by one class rather than another.

But how, when, and why does accumulation by dispossession emerge from this background state to become the dominant form of accumulation relative to expanded reproduction? In part this has to do with how and when crises form in expanded reproduction. But it can also reflect attempts by determined entrepreneurs and developmental states to 'join the system' and seek the benefits of capital accumulation directly.

Any social formation or territory that is brought or inserts itself into the logic of capitalist development must undergo wide-ranging structural, institutional, and legal changes of the sort that Marx describes under the rubric of primitive accumulation. The collapse of the Soviet Union posed exactly this problem. The result was a savage episode of primitive accumulation under the heading of 'shock therapy' as advised by the capitalist powers and international institutions. The social distress was immense, but the distribution of assets that resulted through privatization and market reforms was both lop-sided and not very conducive to the sorts of investment activity that typically emerge with expanded reproduction. Even more recently, the turn towards state-orchestrated capitalism in China has entailed wave after wave of primitive accumulation. Hitherto successful state and township/village enterprises around Shanghai (which provided component parts to major industries in the metropolitan area) have in recent times either been forced to close or be privatized, thus shedding social welfare and pension obligations and creating a huge pool of unemployed and asset-poor workers. The effect has been to make the remaining Chinese enterprises far more fiercely competitive in world markets, but at the expense of the devaluation and destruction of previously viable livelihoods. While accounts remain sketchy, the result seems to have been a great deal of localized social distress and episodes of fierce, sometimes even violent, class struggle in areas desolated by this process.[13]

Accumulation by dispossession can here be interpreted as the necessary cost of making a successful breakthrough into capitalist development with the strong backing of state powers. The motivations can be internally driven (as in the case of China) or externally imposed (as in the case of neo-colonial development in export-processing zones in South-East Asia or the structural reform approach that the Bush administration now proposes to attach to foreign aid grants to poor nations). In most cases, some combination of internal motivation and external pressure lies behind such transformations. Mexico, for example, abandoned its already weakening protections of peasant and indigenous populations in the 1980s, in part under pressure from its neighbour to the north to adopt privatization and neo-liberal practices in return for financial assistance and the opening of the US market for trade through the NAFTA agreement. And even when the motivation appears predominantly internal, the external conditions matter. The setting up of the WTO makes it easier now for China to break into the global capitalist system than would have been the case back in the 1930s when autarky within closed empires prevailed, or even back in the 1960s, when the state-dominated Bretton Woods system kept capital flows under stricter control. Post-1973 conditions—and this has been the obverse of what US pressures to open markets was supposed to do—have been far more favourable for any country or regional complex that wished to insert itself into the global capitalist system—hence the rapid rise of territories such as Singapore, Taiwan, and South Korea, and several other newly industrializing regions and countries. This openness of opportunity brought waves of deindustrialization to much of the advanced capitalist world (and even beyond, as we saw in Chapter 3) at the same time as it rendered the newly industrializing countries, as in the crisis of 1997–8, more vulnerable to movements of speculative capital, spatio-temporal competition, and further waves of accumulation by dispossession. Thus was the volatility of international capitalism constructed and expressed.

The devaluations inflicted in the course of crises are often destructive of social well-being and of social institutions more generally. This typically arises when the credit system operates a squeeze, when liquidity dries up and enterprises are forced into bankruptcy. There is no way for owners to hang on to assets and they have to relinquish them at a very low price to capitalists who have the liquidity to take over. But the circumstances vary widely. The displacements that occurred in the Dust Bowl of the 1930s and the mass migration of the 'okies' to California (so dramatically described in Steinbecks's *Grapes of*

Wrath) was the violent precursor to a long process of gradual displacement of family farming in the United States by agribusiness. The prime lever for this transition has always been the credit system, but perhaps the most interesting aspect of it is how a variety of state institutions set up ostensibly to help preserve family farming played a subversive role in facilitating the transition they were supposed to hold back.

Accumulation by dispossession became increasingly more salient after 1973, in part as compensation for the chronic problems of overaccumulation arising within expanded reproduction. The primary vehicle for this development was financialization and the orchestration, largely at the behest of the United States, of an international financial system that could, from time to time, visit anything from mild to savage bouts of devaluation and accumulation by dispossession on certain sectors or even whole territories. But the opening up of new territories to capitalist development and to capitalistic forms of market behaviour also played a role, as did the primitive accumulations accomplished in those countries (such as South Korea, Taiwan, and now, even more dramatically, China) that sought to insert themselves into global capitalism as active players. For all of this to occur required not only financialization and freer trade, but a radically different approach to how state power, always a major player in accumulation by dispossession, should be deployed. The rise of neo-liberal theory and its associated politics of privatization symbolized much of what this shift was about.

Privatization: The Cutting Edge of Accumulation by Dispossession

Neo-liberalism as a political economic doctrine goes back to the late 1930s. Radically opposed to communism, socialism, and all forms of active government intervention beyond that required to secure private property arrangements, market institutions, and entrepreneurial activity, it began as an isolated and largely ignored corpus of thought that was actively shaped during the 1940s by thinkers such as von Hayek, Ludvig von Mises, Milton Friedman, and, at least for a while, Karl Popper. It would, presciently predicted von Hayek, take at least a generation for neo-liberal views to become mainstream. Assembling funds from sympathetic corporations and founding exclusive think-tanks, the movement produced a steady but ever-expanding stream of analyses, writings, polemics, and political position statements during the 1960s and 1970s. But it was still dismissed as largely irrelevant and even scoffed at by the mainstream. It was only after the general crisis of overaccumulation became so apparent in the 1970s that the movement was taken seriously as an alternative to Keynesian and other more state-centred frameworks for policy-making. And it was Margaret Thatcher who, casting around for a better framework for attacking the economic problems of her time, discovered the movement politically and turned to its think-tanks for inspiration and advice after her election in 1979.[14] Together with Reagan, she transformed the whole orientation of state activity away from the welfare state and towards active support for the 'supply-side' conditions of capital accumulation. The IMF and the World Bank changed their

157

policy frameworks almost overnight, and within a few years neo-liberal doctrine had made a very short and victorious march through the institutions to dominate policy, first in the Anglo-American world but subsequently throughout much of the rest of Europe and the world. Since privatization and liberalization of the market was the mantra of the neo-liberal movement, the effect was to make a new round of 'enclosure of the commons' into an objective of state policies. Assets held by the state or in common were released into the market where overaccumulating capital could invest in them, upgrade them, and speculate in them. New terrains for profitable activity were opened up, and this helped stave off the overaccumulation problem, at least for a while. Once in motion, however, this movement created incredible pressures to find more and more arenas, either at home or abroad, where privatization might be achieved.

In Thatcher's case, the large stock of social housing was one of the first set of assets to be privatized. At first blush this appeared as a gift to the lower classes, who could now convert from rental to ownership at a relatively low cost, gain control over a valuable asset, and augment their wealth. But once the transfer was accomplished housing speculation took over, particularly in prime central locations, eventually bribing, cajoling, or forcing low-income populations out to the periphery in cities like London, and turning erstwhile working-class housing estates into centres of intense gentrification. The loss of affordable housing produced homelessness and social anomie in many urban neighbourhoods. In Britain, the subsequent privatization of utilities (water, telecommunications, electricity, energy, transportation), the selling off of any publicly owned companies, and the shaping of many other public institutions (such as universities) according to an entrepreneurial logic meant a radical transformation in the dominant pattern of social relations and a redistribution of assets that increasingly favoured the upper rather than the lower classes.

The same pattern of asset redistribution can be found almost anywhere that privatization occurred. The World Bank treated post-apartheid South Africa as a showcase for the greater efficiencies that could be achieved through privatization and liberalization of the market. It promoted, for example, either the privatization of water or 'total cost recovery' by municipally owned utilities. Consumers paid for the water they used, rather than receiving it as a free good. With higher revenues the utilities would, the theory went, earn profits and extend services. But, unable to afford the charges, more and more people were cut out of the service, and with less revenue the companies raised rates, making water even less affordable to low-income populations. One outcome, as they were forced to turn to other sources of water supply, was a cholera epidemic in which many people died. The stated objective (running water for all) could not be realized given the means insisted upon. Extensive surveys in South Africa by McDonald and others thus show that 'cost recovery on municipal services imposes enormous hardships on low-income families, contributes to massive numbers of service cut-offs and evictions, and jeopardises the potential for millions of low-income families to lead healthy and productive lives'.[15]

This same logic took Argentina through an extraordinary wave of privatization (water, energy, telecommunications, transportation) which resulted in a huge inflow of over-accumulated capital and a substantial boom in asset values, followed by a collapse into massive impoverishment (now extended to more than half of the population) as capital withdrew to go elsewhere. Consider, as another example, the case of Mexican land rights. The 1917 Constitution from the Mexican revolution protected the legal rights of indigenous peoples and enshrined those rights in the *ejido* system, which allowed land to be collectively held and used. In 1991 the Salinas government passed a reform law that both permitted and encouraged privatization of the *ejido* lands. Since the *ejido* provided the basis for collective security among indigenous groups, the government was, in effect, divesting itself of its responsibilities to maintain the basis for that security. This was, moreover, one item within a general package of privatization moves under Salinas which dismantled social security protections in general and which had predictable and dramatic impacts upon income and wealth distribution.[16] Resistance to the *ejido* reform was widespread, and the most vociferous of the campesino groups ended up supporting the Zapatista rebellion that broke out in Chiapas on the very day in January 1994 when the NAFTA accord was due to be put into effect. The subsequent lowering of import barriers delivered yet another blow as cheap imports from the efficient but also highly subsidized agribusinesses (as much as 20 per cent of cost) in the United States drove down the price of corn and other products to the point where small agricultural producers could not compete. Close to starvation, many of these producers have been forced off the land to augment the pool of the unemployed in already overcrowded cities. Similar effects on rural populations have been experienced world-

160

wide. Cheap imports of vegetables from California and rice from Louisiana, achieved under WTO rules, are now displacing rural populations in Japan and Taiwan for example. Foreign competition under WTO rules is devastating rural life in India. In effect, reports Roy, 'India's rural economy, which supports seven hundred million people, is being garroted. Farmers who produce too much are in distress, farmers who produce too little are in distress, and landless agricultural labourers are out of work as big estates and farms lay off their workers. They're all flocking to the cities in search of employment.'[17] In China the estimate is that at least half a billion people will have to be absorbed by urbanization over the next ten years if rural mayhem and revolt is to be avoided. What they will do in the cities remains unclear, though, as we have seen, the vast physical infrastructural plans now in the works will go some way to absorbing the social distress.

Privatization, Roy concludes, is essentially 'the transfer of productive public assets from the state to private companies. Productive assets include natural resources. Earth, forest, water, air. These are the assets that the state holds in trust for the people it represents. . . . To snatch these away and sell them as stock to private companies is a process of barbaric dispossession on a scale that has no parallel in history.'[18]

That the Zapatista rebellion in Chiapas, Mexico had much to do with protection of indigenous rights was obvious. That the trigger for this movement was the conjoining of initiatives towards privatization of the commons and the opening up of free trade through NAFTA was also obvious. This raises, however, the general question of the resistance to accumulation by dispossession.

Struggles over Accumulation by Dispossession

Primitive accumulation as Marx depicts it entailed a whole series of violent and episodic struggles. The birth of capital was no peaceable affair. It was written into the history of the world, as Marx put it, 'in letters of blood and fire'. Christopher Hill, in *The World Turned Upside Down*, provides a detailed account of how these struggles unfolded in seventeenth-century Britain, as the forces of private power and landownership clashed repeatedly with multiple and diverse popular movements pointing away from capitalism and privatization towards radically different forms of social and communal organization.[19] Accumulation by dispossession in our own times has similarly provoked political and social struggles and vast swaths of resistance. Many of these now form the core of a diverse and seemingly inchoate but widespread anti- or alternative globalization movement. The ferment of alternative ideas within these movements matches the fecundity of ideas generated in other historical phases of parallel disruptions in ways of life and social relations (1640–80 in Britain and 1830–48 in France spring to mind). The emphasis within these movements on the theme of 'reclaiming the commons' is indicative, however, of the deep continuities with struggles of long ago.

These struggles pose, however, serious difficulties of interpretation and analysis. You cannot make an omelette without breaking eggs, the old adage goes, and the birth of capitalism entailed fierce and often violent episodes of creative destruction. While the class violence was abhorrent, the positive side was to obliterate feudal relations, liberate creative energies, open up society to strong currents of technological and organizational change, and overcome a world based on superstition and ignorance and replace it with a world of scientific enlightenment with the potentiality to liberate people from material want and need. From this standpoint it could be said that primitive accumulation was a necessary though ugly stage through which the social order had to go in order to arrive at a state where both capitalism and some alternative socialism might be possible. Marx (as opposed to anarchists such as Reclus and Kropotkin, and adherents of the William Morris branch of socialism) placed little if any value on the social forms destroyed by primitive accumulation. Nor did he argue for a perpetuation of the status quo and most certainly not for any reversion to precapitalist social relations and productive forms. He took the view that there was something progressive about capitalist development and that this was true even for British imperialism in India (a position that did not command much respect in the anti-imperialist movements of the post-Second World War period, as the icy reception of Bill Warren's work on imperialism as the pioneer of capitalism showed).[20]

This issue is of critical importance in any political evaluation of contemporary imperialistic practices. While levels of exploitation of labour power in developing countries are undoubtedly high and abundant cases of abusive practices can be identified, the ethnographic accounts of the social transformations wrought by foreign direct investments, industrial development, and offshore production systems in many parts of the world tell a more complicated story. In some instances the position of women, who provide the bulk of the labour power, has been significantly changed if not enhanced. Faced with the choice of sticking with industrial labour or returning to rural impoverishment, many within the new proletariat seem to express a strong preference for the former. In other instances sufficient class power has been achieved to make real material gains in living standards and to achieve a standard of life far superior to the degraded circumstances of a previous rural existence. It is then arguable whether the problem in Indonesia, for example, was the impact of rapid capitalist industrialization on life chances during the 1980s and 1990s or the devaluation and de-industrialization occasioned through the financial crisis of 1997–8 that demolished much of what that industrialization had achieved. Which, then, was the more serious problem: the import and insertion of capital accumulation through expanded reproduction into the Indonesian economy or the total disruption of that activity through accumulation by dispossession? While it is obviously true that the latter was a logical corollary of the former, and that the real tragedy is constituted by drawing (sometimes forcibly) populations into the proletariat in short order only to cast them off as redundant labour, I also think it plausible that the second step did far more damage to the long-term hopes, aspirations, and possibilities of the mass of the impoverished population than did the first. The implication is that primitive accumulation that opens up a path to expanded reproduction is one thing, and accumulation by dispossession that disrupts and destroys a path already opened up is quite another.

The recognition that primitive accumulation may be a necessary precursor to more positive changes raises the whole question of the politics of dispossession under

Accumulation by Dispossession

socialism. It was, within the Marxist/communist revolutionary tradition, often deemed necessary to organize the
equivalent of primitive accumulation in order to imple-
ment programmes of modernization in those countries
that had not gone through the initiation into capitalist
development. This sometimes meant similar levels of
appalling violence, as with the forced collectivization of
agriculture in the Soviet Union (the elimination of the
kulaks) and in China and eastern Europe. These policies
were hardly great success stories and sparked political
resistance that was in some instances ruthlessly crushed.
This approach created its own difficulties wherever it
was implemented. The difficulties the Sandinistas had in
dealing with the Atlantic coast Mesquito Indians in
Nicaragua, as they planned socialist development in the
region, created a Trojan horse through which the CIA
could mount its successful Contra offensive against the
revolution.

While, therefore, struggles against primitive accumulation could provide the seedbed of discontent for insurgent
movements, including those embedded in the peasantry,
the point of socialist politics was not to protect the ancient
order but to attack directly the class relations and forms of
state power that were attempting to transform it and
arrive thereby at a totally different configuration of class
relations and state powers. This idea was central to many
of the revolutionary movements that swept the developing
world in the aftermath of the Second World War. They
fought against capitalist imperialism but did so in the
name of an alternative modernity rather than in defence of
tradition. In so doing they often found themselves oppos-
ing and opposed by those who sought to protect if not

revitalize traditional systems of production, cultural norms, and social relations.

Insurgent movements against accumulation by dispossession did not necessarily appreciate being co-opted by socialist developmentalism. The patchy record of success for the socialist alternative (the early achievements of Cuba in fields of health care, education, and agronomy initially inspired before later flagging), and the climate of repressive politics largely orchestrated by Cold War politics, made it increasingly difficult for the traditional left to claim a position of leadership rather than of coercive domination in relation to these social movements.

The insurgent movements against accumulation by dispossession generally took a different political path, in some instances quite hostile to socialist politics. This was sometimes for ideological but in other instances simply for pragmatic and organizational reasons that derived from the very nature of what such struggles were and are about. To begin with, the variety of such struggles was and is simply stunning. It is hard to even imagine connections between them. The struggles of the Ogoni people against the degradation of their lands by Shell Oil; the long-drawn-out struggles against World Bank-backed dam construction projects in India and Latin America; peasant movements against biopiracy; struggles against genetically modified foods and for the authenticity of local production systems; fights to preserve access for indigenous populations to forest reserves while curbing the activities of the timber companies; political struggles against privatization; movements to procure labour rights or women's rights in developing countries; campaigns to protect biodiversity and to prevent habitat destruction; peasant movements to gain access to land; protests against high-way and airport construction; literally hundreds of protests against IMF-imposed austerity programmes—these were all part of a volatile mix of protest movements that swept the world and increasingly grabbed the head-lines during and after the 1980s.[21] These movements and revolts were frequently crushed with ferocious violence, for the most part by state powers acting in the name of 'order and stability'. Client states, supported militarily or in some instances with special forces trained by the major military apparatuses (led by the US, with Britain and France playing a minor role), took the lead in a system of repressions and liquidations to ruthlessly check activist movements challenging accumulation by dispossession.

To this complicated picture must then be added the extraordinary proliferation of international NGOs, par-ticularly after 1970 or so, most of them dedicated to sin-gle-issue politics (the environment, the status of women, civil rights, labour rights, poverty elimination, and the like). While some of these NGOs came out of religious and humanistic traditions in the West, others were set up in the name of battling poverty but were funded by groups assiduously pursuing the aim of proliferating market exchange. It is hard not to feel overwhelmed by the extent and diversity of issues or the range of objec-tives. An activist like Roy puts it this way: 'What is hap-pening to our world is almost too colossal for human comprehension to contain. But it is a terrible, terrible thing. To contemplate its girth and circumference, to attempt to define it, to try and fight it all at once, is impossible. The only way to fight it is by fighting specific wars in specific ways.'[22]

But the movements are not only inchoate. They often exhibit internal contradictions as, for example, when indigenous populations claim back rights in areas that environmental groups regard as crucial to put a fence around to protect biodiversity and to prevent habitat destruction. And partly because of the distinctive conditions that give rise to such movements, their political orientation and modes of organization also depart markedly from those that typically coalesced around expanded reproduction. The Zapatista rebellion, for example, did not seek to take over state power or accomplish a political revolution. It sought instead a more inclusionary politics to work through the whole of civil society in a more open and fluid search for alternatives that would look to the specific needs of the different social groups and allow them to improve their lot. Organizationally, it tended to avoid avant-gardism and refused to take on the form of a political party. It preferred instead to remain a social movement within the state, attempting to form a political power bloc in which indigenous cultures would be central rather than peripheral. It sought thereby to accomplish something akin to a passive revolution within the territorial logic of power commanded by the Mexican state apparatus.[23]

The effect of all these movements *in toto* was to shift the terrain of political organization away from traditional political parties and labour organizing into what was bound to be in aggregate a less focused political dynamic of social action across the whole spectrum of civil society. What this movement lost in focus it gained in terms of relevance and embeddedness in the politics of daily life. It drew its strengths from that embeddedness, but in so doing often found it hard to extract itself from the local and the particular to understand the macro–politics of what accumulation by dispossession was and is all about.

The danger, however, is of seeing all such struggles against dispossession as by definition 'progressive' or, even worse, of placing them under some homogenizing banner like that of Hardt and Negri's 'multitude' that will magically rise up to inherit the earth.[24] This, I think, is where the real political difficulty lies. Because if Marx is only partially right, in holding that there can sometimes be something progressive about primitive accumulation, that to make the omelette some eggs must be broken, then we have to confront difficult choices head–on. And these are the choices that now face the anti- or alternative glob-alization movement and which threaten to blow apart a movement that seems to hold such promise for anti-capitalist and anti-imperialist struggle. Let me elaborate.

The Dual Domains of Anti-Capitalist and Anti-Imperialist Struggle

The classic view of the Marxist/socialist left was that the proletariat, defined as waged workers deprived of access to or ownership of the means of production, was the key agent of historical change. The central contradiction was between capital and labour in and around the point of pro-duction. The primary instruments of working–class organization were trade unions and political parties whose aim was to procure state power in order either to regulate or to supplant capitalist class domination. The focus was, therefore, on class relations and class struggles within the field of capital accumulation understood as expanded reproduction. All other forms of struggle were viewed as subsidiary, secondary, or even dismissed as peripheral or irrelevant. There were, of course, many nuances and variations on this theme but at the heart of it all the view prevailed that the proletariat was the unique agent of historical transformation. Struggles waged according to this prescription bore remarkable fruit for much of the twentieth century, particularly in the advanced capitalist countries. While revolutionary transformations did not occur, the growing power of working-class organizations and political parties achieved remarkable improvements in material living standards coupled with the institutionalization of a wide range of social protections. The social democratic welfare states that emerged, particularly in western Europe and Scandinavia, could be viewed, in spite of their inherent problems and difficulties, as models of progressive development. And they would not have come about had it not been for fairly single-minded proletarian organization within the framework of expanded reproduction as experienced within the nation-state. I think it important to acknowledge the significance of this achievement.

While the single-mindedness was productive, it was bought at the cost of innumerable exclusions. Attempts, for example, to incorporate urban social movements into the agenda of the left broadly failed, except, of course, in those parts of the world where communitarian politics prevailed. The politics deriving from the workplace and the point of production dominated the politics of the living space. Social movements such as feminism and environmentalism remained outside the purview of the traditional left. And the relation of internal struggles for social betterment to external displacements characteristic of imperialism tended to be ignored (with the result that much of the labour movement in the advanced capitalist countries fell into the trap of acting as an aristocracy of labour out to preserve its own privileges, by imperialism if necessary). Struggles against accumulation by dispossession were considered irrelevant. This single-minded concentration of much of the Marxist- and communist-inspired left on proletarian struggles to the exclusion of all else was a fatal mistake. For if the two forms of struggle are organically linked within the historical geography of capitalism, then the left was not only disempowering itself but was also crippling its analytical and programmatic powers by totally ignoring one side of this duality.

In the long-drawn-out dynamic of class struggle after the crisis of 1973, working-class movements were everywhere put on the defensive. While there was considerable unevenness in how these struggles unfolded (depending upon the strength of resistance), the effect was generally to diminish the power of these movements to affect the trajectory of global capitalist development. The rapid expansion of production in East and South-East Asia occurred in a world where, with the single exception of South Korea, independent (as opposed to corporatist) trade-union movements were either non-existent or vigorously repressed and where communism and socialism as political movements were violently put down (the Indonesian bloodbath of 1965, when Suharto overthrew Sukarno and maybe as many as a million people were killed, was the most brutal case). Elsewhere, throughout Latin America as well as in Europe and North America, the rise of finance capital, freer trade, and the disciplining of the state by cross-border flows in liberalized capital markets made traditional forms of labour organization less appropriate and, as a consequence, less successful. Revolutionary and even reformist movements (as in Chile under Allende) were violently repressed by military power.

But the intense difficulty of sustaining expanded reproduction was also generating a much greater emphasis upon a politics of accumulation by dispossession. The forms of organization developed to combat the former did not translate well when it came to confronting the latter. Generalizing crudely, the forms of left-wing political organization established in the period 1945–73, when expanded reproduction was in the ascendant, were inappropriate to the post-1973 world, where accumulation by dispossession moved to the fore as the primary contradiction within the imperialist organization of capital accumulation.

The result was the emergence of a different kind of politics of resistance, armed, eventually, with a different kind of alternative vision to that of socialism or communism. This distinction was early recognized by, for example, Samir Amin, specifically with respect to struggles in what he termed the peripheral zones of capitalism:

the unequal development immanent in capitalist expansion has placed on the agenda of history another type of revolution, that of the peoples (i.e. not specific classes) of the periphery. This revolution is anti-capitalist in the sense that it is against capitalist development as it actually exists because it is intolerable for these peoples. But that does not mean that these anti-capitalist revolutions are socialist. . . . By force of circumstances, they have a complex nature. The expression of their specific and new contradictions, which was not imagined in the classical perspective of the socialist transition as conceived by Marx, gives post-capitalist regimes their real content, which is that of a popular national construction in which the three tendencies of socialism, capitalism and statism combine and conflict.

Unfortunately, Amin went on to argue, many contemporary movements

feed on the spontaneous popular revolt against the unacceptable conditions created by peripheral capitalism; they have so far, however, fallen short of making the demand for the double revolution by which modernization and popular enfranchisement must come together; as a result, their fundamental dimension, feeding on the backward-looking myth, continues to express itself in a language in which the metaphysical concern remains exclusive in the whole social vision.[25]

While I do not think that accumulation by dispossession is exclusively to the periphery, it is certainly the case that some of its most vicious and inhumane manifestations are in the most vulnerable and degraded regions within uneven geographical development.

Struggles over dispossession occur, however, on a variety of scales. Many are local, others regional, and still others global, so that command of the state apparatus—the primary objective of traditional socialist and communist movements—seems less and less relevant. When this shift is coupled with a growing sense of disillusion with what socialist developmentalism has been able to accomplish, then the grounds for seeking an alternative politics appear even stronger. The targets and objectives of such struggles are also, as Amin remarks, diffuse, very much a function of the inchoate, fragmentary, and contingent forms taken by accumulation by dispossession. Destruction of habitat here, privatization of services there, expulsions from the land somewhere else, biopiracy in yet another realm—each creates its own dynamic. The trend is, therefore, to look to the ad hoc but more flexible organizational forms that can be built within civil society to respond to such struggles. The whole field of anti-capitalist, anti-imperialist, and anti-globalization struggle has consequently been reconfigured and a very different political dynamic has been set in motion.

For many commentators, these new movements with their special qualities earned the appellation 'postmodern'. This was how the Zapatista rebellion was often characterized. While the descriptions of such movements were undoubtedly apt, the appellation 'postmodern' is unfortunate. It may seem silly to quarrel about a word, but the substantive connotations are important. There is, to begin with, a certain difficulty that arises out of the inherent periodization and historicism that inevitably attaches to the prefix 'post'. There have been, as I have already indicated, many episodes of primitive accumulation and accumulation by dispossession within the historical geography of capitalism. Eric Wolf's study *Peasant Wars of the Twentieth Century* puts one dimension of such struggles in a comparative perspective without in any way resorting to the idea of postmodernity. It is therefore somewhat surprising to find June Nash, whose depictions of the changing state of things in Chiapas provides an evidentiary document of an exemplary sort, agreeing to the appellation of 'postmodern' for what the Zapatistas were and are about, when it surely makes more sense to see that struggle against the background of a long lineage of such struggles on the part of indigenous and peasant populations against the encroachments of capitalist imperialism and the constant threat of dispossession of whatever assets they do control by state-led actions. In the Zapatista case it is, I think, particularly significant that the struggle first emerged in the lowland forests, where displaced indigenous elements constructed an alliance with *mestizos* based upon their parallel impoverishment and their systematic exclusion from any of the benefits to be had from resource extraction (primarily of oil and timber) from the region they inhabited. The subsequent depiction of this movement as being purely about 'indigenous peoples' may have had more to do with claiming legitimacy with respect to the Mexican Constitution's provision protecting indigenous rights than with an actual description of origins.[26]

But in the same way that dismissal of the 'organic link' between accumulation by dispossession and expanded reproduction disempowered and limited the vision of the traditional left, so resort to the conception of postmodern struggle has the same impact upon the newly emerging movements against accumulation by dispossession. Hostility between the two trains of thought and style of organizing is already much in evidence within the anti-globalization movement. A whole wing of it sees the struggle to command the state apparatus as not only irrelevant but an illusory diversion. The answer lies, they say, in localization of everything.[27] That wing likewise tends to dismiss the union movement as a closed modernist, reactionary, and oppressive form of organization that needs to be superseded by the more fluid and open

175

postmodern forms of social movement. The nascent union movements in, say, Indonesia and Thailand, which are struggling against exactly the same neo-liberal forces of oppression as the Zapatistas, though under very different circumstances and from a very different social and cultural base, find themselves excluded. On the other hand, many traditional socialists regard the new movements as naive and self-destructive, as if there is nothing of interest to be learned from them. Cleavages of this sort are divisive, as some of the debates in the most recent World Social Forums at Porto Alegre have indicated. The accession of the Brazilian Workers' Party, which obviously has a 'workerist' base and seeks to command support in part by traditional leftist means, to state power renders the debate both more strident and more urgent.

But the differences cannot be buried under some nebulous concept of 'the multitude' in motion either. They must be confronted politically as well as analytically. On the latter plane, Luxemburg's formulation stands as extremely helpful. Capital accumulation indeed has a dual character. But the two aspects of expanded reproduction and accumulation by dispossession are organically linked, dialectically intertwined. It therefore follows that the struggles within the field of expanded reproduction (that the traditional left placed so much emphasis upon) have to be seen in a dialectical relation with the struggles against accumulation by dispossession that the social movements coalescing within the anti- and alternative globalization movements are primarily focusing upon. If the current period has seen a shift in emphasis from accumulation through expanded reproduction to accumulation through dispossession, and if the latter lies at the heart of imperialist practices, then it follows that the balance of interest within the anti- and alternative globalization movement must acknowledge accumulation by dispossession as the primary contradiction to be confronted. But it ought never to do so by ignoring the dialectical relation to struggles in the field of expanded reproduction.

But this then re-poses the problem that not all struggles against dispossession are equally progressive. Just consider the militia movement in the United States, or anti-immigrant sentiments in ethnic enclaves fighting against 'foreign' incursions on what they regard as ancient and venerable land rights. The danger lurks that a politics of nostalgia for that which has been lost will supersede the search for ways to better meet the material needs of impoverished and repressed populations; that the exclusionary politics of the local will dominate the need to build an alternative globalization at a variety of geographical scales; that reversion to older patterns of social relations and systems of production will be posited as a solution in a world that has moved on. There appear to be no easy answers to such questions.

Yet it is often relatively easy to effect some level of reconciliation. Consider, for example, Roy's arguments against the massive investments in dam construction in the Narmada valley in India. Roy favours the provision of cheap electricity to impoverished rural populations. She is not an anti-modernist. Her argument against the dams is: (a) the electricity is expensive relative to other forms of generation while the agricultural benefits (rarely measured) from irrigation appear to be minimal; (b) the environmental costs appear to be huge (again, there is no serious attempt to assess let alone measure them); (c) the vast amount of money flowing into the project benefits a small elite of consultants, engineers, construction companies, turbine producers, etc. (many of which are foreign, including the infamous Enron), and that this money could be much better spent elsewhere; (*d*) all the risk is borne by the state since the participating companies are guaranteed a rate of return; and (*e*) that the hundreds of thousands of people displaced from their lands, their histories, and their livelihoods are mostly either indigenous or marginalized (*dalit*) populations that receive absolutely no compensation and no benefits from the projects. They were not even consulted or informed, and ended up standing waist-deep in water in their villages as the government suddenly filled the dam in one monsoon season. While this is, clearly, a specific war in a particular place that needs to be fought in specific ways, its general class character is clear enough, as is the 'barbaric' process of dispossession.[28] That as many as 30 million people have been displaced by dam projects in India alone over the last fifty years testifies to both the extent and brutality of the process. But the reconciliation depends crucially on recognizing the fundamental political role of accumulation by dispossession as a fulcrum of what class struggle is and should be construed to be about.

My own view, for what it is worth, is that the political movements, if they are to have any macro and long-run impact, must rise above nostalgia for that which has been lost and likewise be prepared to recognize the positive gains to be had from the transfers of assets that can be achieved through limited forms of dispossession (as, for example, through land reform or new structures of decision-making such as joint forest management). They must likewise seek to discriminate between progressive and regressive aspects of accumulation by dispossession and seek to guide the former towards a more generalized political goal that has more universal valency than the many local movements, which often refuse to abandon their own particularity. In so doing, however, ways must be found to acknowledge the significance of the multiple identifications (based on class, gender, locality, culture, etc.) that exist within populations, the traces of history and tradition that arise from the ways in which they made themselves in response to capitalist incursions, as they see themselves as social beings with distinctive and often contradictory qualities and aspirations. Otherwise there is the danger of re-creating the lacunae in Marx's account of primitive accumulation and failing to see the creative potential that resides in what some regard dismissively as 'traditional' and non-capitalistic social relations and systems of production. Some way must be found, both theoretically and politically, to move beyond the amorphous concept of 'the multitude' without falling into the trap of 'my community, locality, or social group right or wrong'. Above all, the connectivity between struggles within expanded reproduction and against accumulation by dispossession must assiduously be cultivated. Fortunately, in this, the umbilical cord between the two forms of struggle that lies in financial institutional arrangements backed by state powers (as embedded in and symbolized by the IMF and the WTO) has been clearly recognized. They have quite rightly become the main focus of the protest movements. With the core of the political problem so clearly recognized, it should be possible to build outwards into a broader politics of creative destruction mobilized against the dominant regime of neo-liberal imperialism foisted upon the world by the hegemonic capitalist powers.

Imperialism as Accumulation by Dispossession

When Joseph Chamberlain led Britain into the Boer War through the annexation of the Witwatersrand at the beginning of the twentieth century, it was clear that the gold and diamond resources were the prime motivation. Yet, as we earlier saw, his conversion to an imperialist logic arose out of the inability to find any internal solutions to the chronic problem of overaccumulation of capital within Britain. This inability had everything to do with the internal class structure that blocked any large-scale application of surplus capitals towards social reform and infrastructural investments at home. The drive of the Bush administration to intervene militarily in the Middle East likewise has much to do with procuring firmer control over Middle Eastern oil resources. The need to exert that control had ratcheted steadily upwards since President Carter first enunciated the doctrine that the United States was prepared to use military means to ensure the uninterrupted flow of Middle Eastern oil into the global economy. Since recessions in the global economy correlate with oil price hikes, so the general lowering of oil prices can be seen as one tactic in seeking to confront the chronic problems of overaccumulation that have arisen over the past three decades. As occurred in Britain at the end of the preceding century, the blockage of internal reform and infrastructural investment by the configuration of class interests during these years has also played a crucial role in the conversion of US politics towards a more and more overt embrace of imperialism. It is tempting, therefore, to see the US invasion of Iraq as the equivalent of Britain's engagement in the Boer War, both occurring at the beginning of the end of hegemony.

But military interventions are the tip of the imperialist iceberg. Hegemonic state power is typically deployed to ensure and promote those external and international institutional arrangements through which the asymmetries of exchange relations can so work as to benefit the hegemonic power. It is through such means that tribute is in effect extracted from the rest of the world. Free trade and open capital markets have become primary means through which to advantage the monopoly powers based in the advanced capitalist countries that already dominate trade, production, services, and finance within the capitalist world. The primary vehicle for accumulation by dispossession, therefore, has been the forcing open of markets throughout the world by institutional pressures exercised through the IMF and the WTO, backed by the power of the United States (and to a lesser extent Europe) to deny access to its own vast market to those countries that refuse to dismantle their protections.

None of this, however, would have assumed the importance it currently does if there had not emerged chronic problems of overaccumulation of capital through expanded reproduction coupled with a political refusal to attempt any solution to these problems by internal reform. The rise in importance of accumulation by dispossession as an answer, symbolized by the rise of an internationalist politics of neo-liberalism and privatization, correlates with the visitation of periodic bouts of predatory devaluation of assets in one part of the world or another. And this seems to be the heart of what contemporary imperialist practice is about. The American bourgeoisie has, in short, rediscovered what the British bourgeoisie discovered in the last three decades of the nineteenth century, that, as Arendt has it, 'the original sin of simple robbery' which made possible the original accumulation of capital 'had eventually to be repeated lest the motor of accumulation suddenly die down'.[29] If this is so, then the 'new imperialism' appears as nothing more than the revisiting of the old, though in a different place and time. Whether or not this is an adequate conceptualization of matters remains to be evaluated.

5

Consent to Coercion

Imperialism of the capitalist sort arises out of a dialectical relation between territorial and capitalistic logics of power. The two logics are distinctive and in no way reducible to each other, but they are tightly interwoven. They may be construed as internal relations of each other. But outcomes can vary substantially over space and time. Each logic throws up contradictions that have to be contained by the other. The endless accumulation of capital, for example, produces periodic crises within the territorial logic because of the need to create a parallel accumulation of political/military power. When political control shifts within the territorial logic, flows of capital must likewise shift to accommodate. States regulate their affairs according to their own distinctive rules and traditions and so produce distinctive styles of governance. A basis is here created for uneven geographical developments, geopolitical struggles, and different forms of imperialist politics. Imperialism cannot be understood, therefore, without first grappling with the theory of the capitalist state in all its diversity. Different states produce different imperialisms, as was obviously so with the British, French, Dutch, Belgian, etc.

imperialisms from 1870 to 1945. Imperialisms, like empires, come in many different shapes and forms. While there may be much that is contingent and accidental— indeed it could not be any other way given the political struggles contained within the territorial logic of power—I believe we can go a long way to establishing a solid inter- pretative framework for the distinctively capitalistic forms of imperialism by invoking a double dialectic of, first, the territorial and capitalist logics of power and, secondly, the inner and outer relations of the capitalist state.

Consider, in this light, the case of the recent shift in form from neo–liberal to neo–conservative imperialism in the United States. The global economy of capitalism underwent a radical reconfiguration in response to the overaccumulation crisis of 1973–5. Financial flows became the primary means of articulating the capitalistic logic of power. But once the Pandora's box of finance capital had been opened, the pressure for adaptive trans- formations in state apparatuses also increased. Step by step many states, led by the United States and Britain, moved to adopt neo–liberal policies. Other states either sought to emulate the leading capitalist powers or were forced to do so through structural adjustment policies imposed by the IMF. The neo–liberal state typically sought to enclose the commons, privatize, and build a framework of open commodity and capital markets. It had to maintain labour discipline and foster 'a good business climate'. If a particular state failed or refused to do so it risked classification as a 'failed' or 'rogue' state. The result was the rise of distinctively neo–liberal forms of imperial- ism. Accumulation by dispossession re-emerged from the shadowy position it had held prior to 1970 to become a major feature within the capitalist logic. In this it did a double duty. On the one hand the release of low-cost assets provided vast fields for the absorption of surplus capitals. On the other, it provided a means to visit the costs of devaluation of surplus capitals upon the weakest and most vulnerable territories and populations. If volatility and innumerable credit and liquidity crises were to be a feature of the global economy, then imperialism had to be about orchestrating these, through institutions like the IMF, to protect the main centres of capital accumulation against devaluation. And this is exactly what the Wall Street–Treasury–IMF complex successfully engaged upon, in alliance with the European and Japanese authorities, for more than two decades.

But the turn to financialization had many internal costs, such as deindustrialization, phases of rapid inflation followed by credit crunches, and chronic structural unemployment. The US for one lost its dominance in production, with the exception of sectors such as defence, energy, and agribusiness. The opening up of global markets in both commodities and capital created openings for other states to insert themselves into the global economy, first as absorbers but then as producers of surplus capitals. They then became competitors on the world stage. What might be called 'sub-imperialisms' arose, not only in Europe but also in East and South-East Asia as each developing centre of capital accumulation sought out systematic spatio-temporal fixes for its own surplus capital by defining territorial spheres of influence. But these spheres of influence were overlapping and interpenetrating rather than exclusive, reflecting the ease and fluidity of capital mobility over space and the networks of spatial interdependency that increasingly ignored state borders.

The benefits of this system were, however, highly concentrated among a restricted class of multinational CEOs, financiers, and rentiers. Some sort of transnational capitalist class emerged that nevertheless focused on Wall Street and other centres such as London and Frankfurt as secure sites for placements of capital. This class looked, as always, to the United States to protect its asset values and the rights of property and ownership across the globe. While economic power seemed to be highly concentrated within the United States, other territorial concentrations of financial power could and did arise. Capital concentrated in European and Japanese markets could take its cut, as could almost any rentier class that positioned itself correctly within the matrix of capitalistic institutions. Debt crises might rock Brazil and Mexico, liquidity crises might destroy the economies of Thailand and Indonesia, but rentier elements within all those countries could not only preserve their capital but actually enhance their own internal class position. Privileged classes could seal themselves off in gilded ghettos in Bombay, São Paulo, and Kuwait while enjoying the fruits of their investments on Wall Street. Just because Wall Street was awash with money did not mean, therefore, that Americans owned that money. Wall Street's problem was to find profitable uses for all the surplus money it commanded, no matter whether it was held by Americans or foreigners.

This geographical dispersal of capitalistic class power did not only apply to rentiers and financial interests; production capital took advantage of the spatial volatility and the shifting territorial logics. The large multinationals in electronics, shoes, and shirts gained remarkably through geographical mobility. But then so did certain other social groups. The Chinese business diaspora, for example, improved its position precisely because it had both the means and the inclination to extract profits out of mobility. Taiwanese and South Korean sub-contractors moved into Latin America and Southern Africa and did extraordinarily well, while those they employed suffered appallingly.[1]

But it was a peculiar feature of this world that an increasingly transnational capitalist class of financiers, CEOs, and rentiers, should look to the territorial hegemon to protect their interests and to build the kind of institutional architecture within which they could gather the wealth of the world unto themselves. This class paid very little heed to place-bound or national loyalties or traditions. It could be multi-racial, multi-ethnic, multicultural, and cosmopolitan. If financial exigencies and the quest for profit required plant closures and the diminution of manufacturing capacity in their own backyard, then so be it. US financial interests were perfectly content to undermine US hegemony in production, for example. This system reached its apogee during the Clinton years, when the Rubin–Summers Treasury Department orchestrated international affairs greatly to the advantage of rentier interests on Wall Street, though they often took very high risks in doing so. The culmination was the disciplining of competition from East and South-East Asia in 1997–8 in such a way as to allow the financial centres of Japan and Europe, but above all the United States, to snap up assets for almost nothing and thereby augment their own profit lines at the cost of massive devaluations and the destruction of livelihoods elsewhere. This was, however, only one example of the innumerable debt and financial crises that afflicted many parts of the developing world after 1980 or so.

Neo-liberal imperialism abroad tended to produce chronic insecurity at home. Many elements in the middle classes took to the defence of territory, nation, and tradition as a way to arm themselves against a predatory neo-liberal capitalism. They sought to mobilize the territorial logic of power to shield them from the effects of predatory capital. The racism and nationalism that had once bound nation-state and empire together re-emerged at the petty bourgeois and working-class level as a weapon to organize against the cosmopolitanism of finance capital. Since blaming the problems on immigrants was a convenient diversion for elite interests, exclusionary politics based on race, ethnicity, and religion flourished, particularly in Europe where neo-fascist movements began to garner considerable popular support. The corporate and financial elites gathered at Davos in 1996 then worried that a 'mounting backlash' against globalization within industrial democracies might have a 'disruptive impact on economic activity and social stability in many countries'. The prevailing mood of 'helplessness and anxiety' was conducive to 'the rise of a new brand of populist politician' and this could 'easily turn into revolt'.[2]

But by then the anti-globalization movement was beginning to emerge, attacking the powers of finance capital and its primary institutions (the IMF and the World Bank), seeking to reclaim the commons, and demanding a space within which national, regional, and local differences could flourish. With the state so clearly siding with the financiers and in any case performing as a prime agent in the politics of accumulation by dispossession, this movement looked to the institutions of civil society to transform the territorial logics of power on a variety of scales, from intensely local to global (as in the case of the environmental movement). The prevalence of fraud, rapine, and violence provoked many violent responses. The surface civilities that supposedly attach to properly functioning markets were little in evidence. The protest movements that surfaced throughout the world were, for the most part, ruthlessly put down by state powers. Low-level warfare raged across the world, often with US covert involvement and military assistance.

Eschewing traditional forms of labour organization, such as unions, political parties, and even the pursuit of state power (now seen as hopelessly compromised), these oppositional movements looked to their own autonomous forms of social organization, even setting up their own unofficial territorial logics of power (as did the Zapatistas), oriented to improving their lot or defending themselves against a predatory capitalism. A burgeoning movement of non-governmental organizations (some of them sponsored by governments) sought to control these social movements and orient them towards particular channels, some of which were revolutionary but others of which were about accommodation to the neo-liberal regime of power. But the result was a ferment of local, dispersed, and highly differentiated social movements battling either to confront or to hold off the neo-liberal practices of imperialism orchestrated by finance capital and neo-liberal states.

The volatility inherent in neo-liberalism ultimately returned to haunt the heartland of the United States itself. The economic collapse that began in the high-tech dot.com economy in 1999 soon spread to reveal that much of what passed for finance capital was in fact unredeemable fictitious capital supported by scandalous accounting practices and totally empty assets. Even before the events of 9/11, it was clear that neo-liberal imperialism was weakening on the inside, that even the asset values on Wall Street could not be protected, and that the days of neo-liberalism and its specific forms of imperialism were numbered. The big issue was what kind of relation between the territorial and capitalistic logics of power would now emerge and what kind of imperialism it would produce.

The fortuitous election of George W. Bush, a born-again Christian, to the US presidency brought a neo-conservative group of thinkers close to power. The neo-conservatives, well funded and organized in numerous 'think-tanks' like the neo-liberals before them, had long sought to impose their agenda on government. And it is a different agenda from that of neo-liberalism. Its primary objective is the establishment of and respect for order, both internally and upon the world stage. This implies strong leadership at the top and unwavering loyalty at the base, coupled with the construction of a hierarchy of power that is both secure and clear. To the neo-conservative movement, adherence to moral principle is also crucial. In this it finds its backbone and electoral base with fundamentalist Christians who hold to beliefs of a very special kind. In the wake of 9/11, for example, Jerry Falwell and Pat Robertson (two major leaders within the movement) expressed the view that the event was a sign of God's anger at the permissiveness of a society that tolerated abortion and homosexuality. Later, on one of the most watched current affairs programmes on American television, Falwell declared that Muhammad was the first great terrorist, while others expressed support for Zionism and for Sharon's violence towards the Palestinians since this would lead to Armageddon and the Second Coming. Belief in the book of Revelation and Armageddon is very widespread (Reagan espoused it, for example). It is hard for Europeans in particular to understand that around a third of the US population holds firmly to such beliefs (including creationism rather than evolution), which imply acceptance of the horrors of war (particularly in the Middle East) as a prelude to the achievement of God's will on earth. Much of the US military is now recruited from the south, where these views are prevalent.

While the neo-conservatives know they cannot stay in power holding to such a platform, the influence of the Christian right cannot be underestimated. The failure to place any constraints on Sharon's violent repression of the Palestinians (interpreted by fundamentalists as a positive step towards Armageddon) is a case in point. And in the conflict with the Arab world it is hard not to let these attitudes slip into the rhetoric of a Christian crusade versus an Islamic jihad, thus converting Huntington's unconvincing thesis of an imminent clash of civilizations into a geopolitical fact.[3]

The neo-conservative charter for foreign policy was laid out in *The Project for the New American Century* that got under way in 1997.[4] The title speaks, as did Luce back in 1941, of a century rather than of territorial control. It deliberately repeats, therefore, all the evasions that Smith exposes in Luce's presentation.[5] The Project is 'dedicated

to a few fundamental propositions: that American leadership is good both for America and for the world; that such
leadership requires military strength, diplomatic energy,
and commitment to moral principle; and that too few
political leaders today are making the case for global leadership'. The principles involved were clearly laid out in
Bush's statement on the anniversary of 9/11 (cited in
Chapter 1 above). Though recognized as distinctive
American values, these principles are presented as univer-
sals, with terms like freedom and democracy and respect
for private property, the individual, and the law bundled
together as a code of conduct for the whole world. The
Project also seeks to 'rally support for a vigorous and prin-
cipled policy of American international involvement'.
This means exporting and if necessary imposing appropriate codes of conduct upon the rest of the world. Most
of the core members of the Project came, however, from
the defence establishment of the former Reagan and Bush
administrations. They are key representatives of that
'military-industrial complex' against whose power
Eisenhower had long ago so clearly warned and which had
grown so much more powerful in the Reagan years. Most
of them joined the new Bush administration. Whereas the
key positions in the Clinton administration were in the
Treasury (where Rubin and Summers ruled supreme),
the new Bush administration looks to its defence
experts—Cheney, Rumsfeld, Wolfowitz, and Powell—to
shape international policy, and relies upon a Christian
conservative—Ashcroft—as Attorney General to enforce
order at home. The Bush administration is, therefore,
dominated by neo-conservatives, deeply indebted to the
military-industrial complex (and a few other major sectors

of American industry, such as energy and agribusiness), and supported in its moral judgements by fundamentalist Christians. Their task was to consolidate power behind a minority-led political agenda within the territorial logic of power. In this they well understood the connection between internal and external order. They intuitively accepted Arendt's view that empire abroad entails tyranny at home, but state it differently. Military activity abroad requires military-like discipline at home.

Iraq had long been a central concern for the neo-conservatives, but the difficulty was that public support for military intervention was unlikely to materialize without some catastrophic event 'on the scale of Pearl Harbor', as they put it. 9/11 provided the golden opportunity, and a moment of social solidarity and patriotism was seized upon to construct an American nationalism that could provide the basis for a different form of imperialist endeavour and internal control. Most liberals, even those who had formerly been critical of US imperialist practices, backed the administration in launching its war against terror and were prepared to sacrifice something of civil liberties in the cause of national security. The accusation of being unpatriotic was used to suppress critical engagement or meaningful dissent. The media and the political parties fell into line. This enabled the political leadership to enact repressive legislation with scarcely any opposition—most notably the Patriot and Homeland Security Acts. Draconian curbs on civil rights were instituted. Prisoners were held illegally and without representation in Guantanamo Bay, indiscriminate round-ups of 'suspects' occurred, and many were held for months without access to legal advice, let alone a trial. Police could arbitrarily detain anyone suspected of 'terrorism', which could include, it soon became clear, even those in the anti-globalization movement. Draconian surveillance techniques were introduced (the FBI was to have access to records of book-borrowing from libraries, book purchases, internet connections, records of student enrolment, membership of scuba-diving clubs, etc.). The administration also seized the opportunity to cut all kinds of programmes for the poor (in the name of sacrifice for a national cause). It imposed a tax-cut programme that grossly favoured the wealthiest 1 per cent of the population (in the name of stimulating the economy) and even proposed the elimination of taxes on dividends in the vain hope that this might bolster asset values on Wall Street. But such policies, coupled with flagrant violations of the Bill of Rights and of American constitutionality, could only be sustained, as Washington, Madison, and many others had long ago recognized and feared, through foreign entanglements of an imperialist sort. Given the threats implied in the events of 9/11, and the climate of suppression of dissent, even liberal opinion swung behind the idea of the invasion of Afghanistan, the routing of the Taliban, and the global hunt for al Qaeda.

To sustain the momentum and realize their ambitions, the paranoid style of American politics had to be put to work. The neo-conservatives had long dwelt on the threats posed by Iraq, Iran, and North Korea, and several other so-called 'rogue states', to the global order. Behind this, however, there always lurked the figure of China, long feared as both unpredictable and potentially a powerful competitor on the world stage. The alliance between the neo-conservatives and the military-industrial complex had pressured Clinton during the 1990s to increase military expenditures and be prepared to fight two regional wars—against, for example, 'rogue states' such as Iraq and North Korea—simultaneously. Iraq was central, in part because of its geopolitical position and dictatorial regime, which was immune to financial disciplining because of its oil wealth, but also because it threatened to lead a secular pan-Arab movement that might dominate the whole of the Middle Eastern region and be able to hold the global economy hostage to its powers over the flow of oil. President Carter, recall, had insisted that any attempt to use oil in this way would not be tolerated, and direct US military commitment to the region dates back to at least 1980. The first Gulf War did not produce regime change in Baghdad, in part because there was no UN mandate for it. The settlement imposed on Iraq was unsatisfactory to both sides. The Iraqis baulked and sanctions were imposed, weapons inspectors were sent in and then expelled, the Kurds were protected in an autonomous zone in the north by military threats, and a low-level war continued in the skies above Iraq as the US and Britain jointly patrolled no-fly zones in both the north and the south. Clinton designated Iraq a 'rogue state' and adopted a policy of regime change in Baghdad but restricted the means to covert action and overt economic sanctions which, the neo-conservatives vociferously argued, would not work.

After 9/11, the neo-conservatives had had their 'Pearl Harbor'. The difficulty was that Iraq plainly had no connection with al Qaeda and the fight against terrorism had to take preference. In the invasion of Afghanistan the military tested out much of its new weaponry in the field, almost as a dress rehearsal for what they might do in Iraq and elsewhere. In the process, the US secured a military presence in Uzbekistan and Kyrgyzstan, within striking distance of the Caspian Basin oilfields (where the extent of reserves is still a mystery and where China is battling fiercely to gain a foothold in order to ensure its own supplies to satisfy its rapidly increasing internal demands). Within six months, and with the defeat of the Taliban in Afghanistan behind it, the US administration began to switch its attention to Iraq. By the summer of 2002 it was clear that the US was committed to force regime change on Baghdad militarily no matter what. The only interesting question was how this would be justified to the American public and internationally. From this point on, the administration resorted to all manner of smokescreens, shifting rhetoric daily, putting out undocumented assertions as if they were proven facts (of the sort described in Chapter 1). It sought to construct a coalition of the willing in which Britain, since it was already heavily involved in daily military action in Iraq (and from which it would have been very difficult to extricate itself), was to take a leading role. At first the US denied any role to the UN and even asserted it had no need for Congressional approval, but on these points it had to concede somewhat to political pressures both domestically and internationally. But it assiduously cultivated the new-found nationalism that was created after 9/11 and harnessed it to the imperial project of regime change in Iraq as essential for domestic security, at the same time as it used the imperial project to put in place ever tighter internal controls (fuelled by terror alerts and other security fears on the domestic front). Unfortunately, as Arendt again so astutely remarks, the coupling of nationalism with imperialism cannot be accomplished without resort to racism, and the degraded popular image of Arabs and Islam and official policies towards visitors and immigrants from Arab countries are all too indicative of a rising tide of racism in the US that may do untold future damage both internally and internationally.

While the situation is now one of rapid flux, accompanied by the usual smoke and mirrors of official pronouncement, it is nevertheless possible to discern roughly where the neo-conservative imperial project wants to go. I therefore conclude with a synopsis of that direction and an assessment of the forces ranged against it.

The neo-conservatives look to the reconstruction of Iraq along the lines pioneered in Japan and Germany after the Second World War. Iraq will be liberalized for open capitalistic development with the aim of ultimately creating a wealthy consumerist society along Western lines as a model for the rest of the Middle East. The necessary social, institutional, and political infrastructures will be put in place under US administration, but gradually give way to a clientelist Iraqi political administration (preferably as weak as the Japanese liberal party). Iraq will remain demilitarized but be protected by US forces that will remain in the Gulf region.[6] Iraqi oil will be used to finance the reconstruction and pay for some of the cost of the war, and, it is hoped, will be delivered to the markets of the world (conveniently denominated in dollars rather than euros) at a sufficiently low price to spark some kind of recovery in the global economy.

This is not, however, the limit to neo-conservatives' imperial ambition. They have already begun to speak of

Iran (which after the occupation of Iraq will be totally surrounded by the US military and clearly threatened) and have launched accusations against Syria that speak of 'consequences'. So obvious have these remarks become that the British Foreign Secretary thought it important to state categorically that Britain would absolutely refuse to participate in any military action against either Syria or Iran. But the neo-conservative position, as articulated by Secretary of Defense Rumsfeld all along, is that the US does not need Britain to accomplish its objectives and that it will go it alone if necessary. Pressure on both Syria and Iran is mounting, while the US also looks to internal reform in Saudi Arabia both to forestall any attempt at a takeover by Islamicists (this was, after all, bin Laden's primary objective) and to deal with the fact that much of the fundamentalist teaching that has fuelled opposition to the US is supported by the Saudis. Meanwhile, the US has now honed, and experimented with in Iraq, a military capacity named 'shock and awe' which would have the power to simultaneously destroy the hundreds of long-range guns that the North Koreans have targeted on Seoul. When it cares to, it can destroy all of North Korea's military power and nuclear capacity in one twelve-hour strike.

Lurking behind all of this appears to be a certain geopolitical vision. With the occupation of Iraq and the possible reform of Saudi Arabia and some sort of submission on the part of Syria and Iran to superior American military power and presence, the US will have secured a vital strategic bridgehead, as was pointed out in Chapter 2, on the Eurasian land mass that just happens to be the centre of production of the oil that currently fuels (and will continue to fuel for at least the next fifty years) not only the global economy but also every large military machine that dares to oppose that of the United States. This should ensure the continued global dominance of the US for the next fifty years. If the US can consolidate its alliances with east European countries such as Poland and Bulgaria, and (very problematically) with Turkey, down to Iraq and into a pacified Middle East, then it will have an effective presence that slashes a line through the Eurasian land mass, separating western Europe from Russia and China. The US would then be in a military and geostrategic position to control the whole globe militarily and, through oil, economically. This would appear particularly important with respect to any potential challenge from the European Union or, even more important, China, whose resurgence as an economic and military power and potentiality for leadership in Asia appears as a serious threat to the neo-conservatives. The neo-conservatives are, it seems, committed to nothing short of a plan for total domination of the globe.[7] In that ordered world of a Pax Americana, it is hoped that all segments may flourish under the umbrella of free-market capitalism. In the neo-conservative view, the rest of the world (or at least all property-owning classes) should and will be grateful for the space allowed for economic development under free-market capitalism everywhere.

The big and open question is, of course, can or will such a project work? There are, doubtless, members even of the Bush administration, as well as in the military, who are not only unconvinced of its feasibility but who may well actively oppose it. The internal balance of forces within the administration currently lies with the neo-conservative bloc but it may not remain so. Much will depend, for example, on whether or not the reputation of the neo-conservatives comes out of the military action in Iraq enhanced or besmirched. A long-drawn-out and messy occupation of Baghdad will have serious consequences for the doctrine that this is a battle for liberation rather than occupation of Iraq, for example.

But the external forces ranged against neo-conservative imperialism are formidable. To begin with, the more explicit this project becomes the more it will almost certainly force an alliance between Germany, France, Russia, China, and others that is by no means bereft of power. A relatively unified Eurasian power bloc, as Kissinger for one fears (see above, p. 85), will not necessarily lose the struggle when pitted against the US. Furthermore, if the US does press on into Iran or Syria, the British will almost certainly have to abandon their support for what will then be clearly recognized as self-serving US imperialism. Almost certainly those European governments, such as Spain and Italy, that have supported the US against the clear wishes of their peoples will fall, turning Europe into a much more unified power bloc opposed to US plans than is currently the case. And global opposition within the United Nations will also likely become much stronger as the US becomes more and more isolated.

The neo-conservatives have squandered much of the US's capacity for moral leadership, and its capacity to lead by genuine consent is already much diminished. Even its cultural influence appears on the wane. The US had, in effect, to try to buy consent in the United Nations (treating the UN almost as if it were a form of traditional Chicago ward politics). But the failure of Turkey, a member of NATO, to be bought off, even in the face of severe economic distress and the threat of retaliatory consequences, is illustrative of a deeper problem. There is very little real consent to be found anywhere in the world, the closest being that of the British which, in the eyes of its own public, is very shaky. The US has given up on hegemony through consent and resorts more and more to domination through coercion. It has long aspired, as Colin Powell put it, to be 'the big bully on the block' (see above, p. 80), but his assertion that this is acceptable because the US is trusted to do the right thing now lacks credibility. The rising tide of popular global opposition, represented by the remarkable world-wide turnout in anti-war demonstrations on 15 February 2003, is a force to be contended with.

It is a fervently held belief among the neo-conservatives that once they have established order throughout the world and demonstrated its benefits the opposition to their militarism both at the popular level and among governments everywhere will largely dissipate. There is more than a little utopianism in this vision, but even a partial fulfilment of it rests crucially upon the nature of the benefits generated and how they might be distributed. Neo-conservatism overlaps neo-liberalism, however, in the belief that free markets in both commodities and capital contain all that is necessary to deliver freedom and well-being to all and sundry. To the degree that this has been shown to be demonstrably false, all that the neo-conservatives have done is to transform the low-intensity warfare waged under neo-liberalism around the globe into a dramatic confrontation that will supposedly resolve problems once and for all. It will continue a political economy that rests on accumulation by dispossession (the dispossession of Iraqi oil being the most flagrant beginning point) and do absolutely nothing to counter the spiralling inequalities that contemporary forms of capitalism are producing. Indeed, if their tax policies are anything to go by, the neo-conservatives will do everything they can to bolster these inequalities, presumably on the grounds that in the long run rewarding initiative and talent in this way will improve the life of all. From this we can expect an increase rather than a diminution in global struggles against dispossession and an increase rather than a diminution in the ferment that has fuelled the anti- and alternative globalization movements even to the point of electing governments, as with Lula in Brazil, which seek to mitigate if not roll back the terrain upon which neo-liberalism can operate. There is, furthermore, nothing here to check the slide into nationalism and exclusionary politics as a means to defend against neo-liberal predation. With the US itself turning more and more to racism as a means to bridge nationalism and imperialism, this kind of disintegration will be very much harder to hold in check.

Beyond this there is the crucial question of how the neo-conservative imperial project will be received within the Arab and more broadly the Islamic world. In this regard the neo-conservatives are stepping onto peculiarly dangerous terrain. To begin with, any rapprochement with the Arab world will have to rest on an acceptable solution to the Arab–Israeli conflict about which the Bush administration has been almost totally silent, except for occasional promissory noises, usually in response to external pressures (particularly from Britain). The reason for the seeming indifference and refusal to make any attempt to curb Sharon's policies in Israel lies in the unholy alliance of Zionist influences, strongly supported by the fundamentalist Christians for their own eschatological reasons, within the United States. The failure to conjure any Palestinian settlement out of the deployment of American imperial power in the region will be a permanent strike against the US within and even beyond the Arab world. It will undoubtedly be the source of opposition registered as sporadic violence against both Israel and the United States and perhaps spark internal revolutions within the Muslim world. Secondly, the idea that Iraq can serve as a demonstration project to wean the Islamic world away from its own brands of fundamentalism and its anti-democratic ways, rests on the far-fetched if not preposterous proposition that somehow Iraq can be transformed overnight into a prosperous, capitalistic, and democratic state under US tutelage. On this point the choice of Iraq does make some sense, since it is a country that has not only oil wealth but a great deal of scientific talent and technical know-how; it also had, before the US and Saddam jointly destroyed it, a significant manufacturing and agrarian base. Surplus capital would most certainly find an outlet in rebuilding much of that, but given the neo-liberal rules that still broadly regulate trade and financial flows, and the general state of overaccumulation, it is hard to see Iraq becoming the equivalent of South Korea in the next few years. And even if it began to do so, it is not at all clear that any demonstration effects will occur, given the broad developmental failures of those states, such as Pakistan and Egypt, which have sought a path to capitalist-style economic development over the past two decades with a good deal of US support. The only circumstance in which some hope for Iraqi economic development under occupation may rest is through a recovery of the global economy on an even broader scale than that which occurred in the aftermath of the Second World War.

This brings us more critically to the issue of the economic circumstances that now prevail and the degree to which the processes outlined earlier point to a capitalistic logic of power that is in any way consistent with or malleable to the specific territorial logic that neo-conservative imperialism seeks to impose. While, as always, it is hard to predict with any certainty, there appears to be a deep inconsistency if not outright contradiction between the two logics. If that is so, then either the territorial logic or the capitalistic logic will have to give way or face catastrophic consequences. What, then, are the main signs of this disjuncture?

To begin with, there is the cost of the war itself. It cannot be less than $200 billion and will possibly be much more. To be sure, there is plenty of surplus capital to fund it, but it will demand its rate of return, which either means profits of defence and reconstruction contractors and/or payments of interest on government debts. Dropping bombs is not productive investment and returns no value back into the circulation and accumulation process, unless, that is, we consider a fall in the price of oil to $20 a barrel as part of a rate of return on military action in Iraq. Iraqi oil could, of course, be appropriated to pay for the war, but this would largely preclude its use for internal redevelopment and thereby thwart the possibility of Iraq performing the role of demonstration project for capitalistic development. It will, by all accounts, take several years to bring Iraqi oil production up to the level where it might conceivably fund both. And in any case Iraq has past debts of around $200 billion ($64 billion to Russia alone) as well as outstanding claims against it for compensation deriving from the invasion of Kuwait amounting to over $100 billion. If, under US tutelage, Iraq fails to honour these debts, the international uproar will be considerable (with Russia in the vanguard).

There is, therefore, little option except for the US to go heavily into debt to fund the war. The general effects of a soaring budgetary deficit in the United States would not be benign even under the best of circumstances. But under current conditions of economic stagnation, declining asset values, and disappearing tax revenues, such deficit spending for military purposes will likely push the economy even deeper into recession rather than help revive it internally. Military expenditures are sometimes construed (for example, by Luxemburg) as an economic stimulus (sometimes called 'military Keynesianism'), but they can at best operate only in the very short term (about the length of time it takes to replace equipment and *matériel* used up). And in the present conjuncture any short-term stimulus from this direction is totally offset by declining consumer confidence and a climate of fear (used directly by the administration for its own purposes) that inhibits people from travel or engaging in any activity that appears risky. Hence airlines are either close to or in bankruptcy and tourism and leisure activities are in deep economic difficulty. Losses of jobs and of social protections (such as health insurance and even pension funds) are reverberating throughout the US economy. New York

City's economy, for example, is now in an even more parlous state than it was in the crisis of 1973–5 and its budget deficit looks set to push it into technical bankruptcy within a couple of years.

This problem is exacerbated by the parlous international position of the US economy. Foreigners now own over a third of US government debt and 18 per cent of corporate debt (more than double the ratios in around 1980), and the US now depends on over $2 billion a day of net foreign investment inflow to cover its continuously rising current account deficit with the rest of the world.[8] As argued earlier, this renders the US economy extraordinarily vulnerable to capital flight, some signs of which are already to be seen in the fall in the relative value of the dollar in world markets. The tables are in danger of being turned with respect to the powers of finance capital to support, rather than seriously damage, the United States itself. The capitalistic logic, without the effective state action of which the Bush administration appears incapable, points to the draining away of economic power from the United States rather than the powerful inward movement that was orchestrated during the economic boom of the 1990s. In the same way that speculative capital flowed into Thailand, Indonesia, and Argentina to fuel booms that suddenly collapsed into capital flight and economic catastrophe, so the flight of speculative capital to Wall Street in the 1990s generated a boom that can equally well be (and to some degree already is being) reversed. The circumstances are, of course, somewhat different because the dollar has always been the safe haven for global capital and the power of seigniorage still lies with the US. But much depends upon confidence in the

US government, and the more it is recognized that it is currently dominated by a coalition of the military-industrial complex, neo-conservatives, and, even more worryingly, fundamentalist Christians, the more the logic of capital will look to regime change in Washington as necessary to its own survival. This would have the effect of bringing the neo-conservative version of imperialism to a crashing halt. If this does not happen, the vast drain imposed by an even stronger turn to a permanent war economy may amount to a form of economic suicide for the United States. The surge towards militarism will then appear as a last desperate move by the US to preserve its global dominance at all costs.

But there is one other aspect of the potential damage that the neo-conservative imperialist project might inflict. The unilateralist assertion of US imperial power fails entirely to recognize the high degree of cross-territorial integration that now exists within the capitalistic organization of the circulation and accumulation of capital. Threats of US boycotts of French and German goods and reciprocal boycotts by Europeans hardly make sense when the foreign content of goods in any economy typically lies somewhere between a third and a half of their value. But rising nationalism, now as much promoted by the war as by the oppositional movements towards neo-liberalism, can indeed impose constraints on international capital flow and the dynamics of accumulation. Withdrawal into regional configurations of capital circulation and accumulation, signs of which already abound, can be exacerbated by any rising tide of nationalism and racism, to say nothing of the way in which the idea of a clash of civilizations is gaining ground. But withdrawal into regional power blocs exercising exclusionary practices while engaging in inter-bloc competition is exactly the configuration that spawned the crises of global capitalism in the 1930s and 1940s. Lenin will be proven right. And no one, presumably, wants to revisit that, which makes the slow but discernible drift towards such a resolution even more disconcerting.

Continuation of neo-liberal politics at the economic level, as I have already indicated, entails a continuation if not escalation of accumulation by other means, i.e. accumulation by dispossession. The corollary externally must surely be an ever rising tide of global resistance to which the only answer is the repression by state powers of popular movements. This implies the continuation of the low-intensity warfare that has characterized the global economy for the last twenty years or more unless, that is, some way of assuaging the global overaccumulation problem can be arrived at. The only possibility of that, I have argued, is the disruptive, violent, and gargantuan programme of what is in essence a truly primitive form of accumulation in China that will spark a rate of economic growth and public infrastructural development capable of absorbing much of the world's capital surplus. This presumes that this process does not spark a counter-revolution within China. But, if it succeeds, the drawing off of surplus capital into China will be calamitous for the US economy which currently feeds off capital inflows to support its own unproductive consumption, both in the military and in the private sector. The result would be the equivalent of a 'structural adjustment' in the US economy that would entail an unheard-of degree of austerity the likes of which have not been seen since the Great

Depression of the 1930s. In such a situation, the US would be sorely tempted to use its power over oil to hold back China, sparking a geopolitical conflict at the very minimum in central Asia and perhaps spreading into a more global conflict.

The only possible, albeit temporary, answer to this problem within the rules of any capitalistic mode of production is some sort of new 'New Deal' that has a global reach. This means liberating the logic of capital circulation and accumulation from its neo-liberal chains, reformulating state power along much more interventionist and redistributive lines, curbing the speculative powers of finance capital, and decentralizing or democratically controlling the overwhelming power of oligopolies and monopolies (in particular the nefarious influence of the military-industrial complex) to dictate everything from terms of international trade to what we see, read, and hear in the media. The effect will be a return to a more benevolent 'New Deal' imperialism, preferably arrived at through the sort of coalition of capitalist powers that Kautsky long ago envisaged.

Ultra-imperialism of the kind now favoured in Europe has, however, its own negative connotations and consequences. If Robert Cooper, a Blair adviser, is to be believed, it favours the resurrection of nineteenth-century distinctions between civilized, barbarian, and savage states in the guise of postmodern, modern, and premodern states, with the postmoderns, as guardians of civilized collaborative behaviour, expected to induce by direct or indirect means obeisance to universal (read 'Western' and 'bourgeois') norms, and humanistic (read 'capitalistic') practices across the globe. The postmodern, mainly European, states are, from this perspective, not an 'old Europe' at all but way out ahead of the United States, which seems to have some difficulty shedding its modernist ways. The difficulty is that it was classifications of this sort that allowed nineteenth-century liberals like John Stuart Mill to justify keeping India in tutelage and exacting tribute from abroad while praising the principles of representative government in 'civilized' countries such as their own. In the absence of any strong revival of sustained accumulation through expanded reproduction, this European version of liberal imperialism can only move ever deeper into the neo-liberal quagmire of a politics of accumulation by dispossession throughout the world in order to keep the motor of accumulation from stalling. Such an alternative form of collective imperialism will hardly be acceptable to wide swaths of the world's population who have lived through, and in some instances begun to fight back against, accumulation by dispossession and the predatory forms of capitalism associated with it. The liberal ruse that someone like Cooper proposes is, in any case, far too familiar to postcolonial writers to have much traction.[9]

There are, of course, far more radical solutions lurking in the wings, but the construction of a new 'New Deal' led by the United States and Europe, both domestically and internationally, in the face of the overwhelming class forces and special interests ranged against it, is surely enough to fight for in the present conjuncture. And the thought that it might, by adequate pursuit of some long-term spatio-temporal fix, actually assuage the problems of overaccumulation for at least a few years and diminish the need to accumulate by dispossession might encourage democratic, progressive, and humane forces to align behind it and turn it into some kind of practical reality. This does seem to propose a far less violent and far more benevolent imperial trajectory than the raw militaristic imperialism currently offered up by the neo-conservative movement in the United States.

The real battleground where this has to be fought out, of course, is within the United States. On this count there is some ground for faint hope since the severe curtailment of civil liberties and the long-standing recognition that imperialism abroad will be bought at the cost of tyranny at home provides a serious basis for political resistance, at least on the part of those who truly believe in the Bill of Rights and whose vision of constitutionality is of a different sort to that of the neo-conservative majority that now dominates the Supreme Court. Such people are at least as numerous as the Christian fundamentalists who now wield such a sinister influence in government. And there are signs within the Christian majority, particularly among the leadership (which has broadly articulated an anti-war position), that there is a moral imperative to isolate Christian fundamentalism and to assert a different kind of Christianity that espouses religious tolerance and peaceful coexistence with others. There is an anti-war and anti-imperialist movement struggling to express itself, but the climate of nationalism, patriotism, and suppression of dissent at all levels, particularly within the media, means that there is a daunting struggle to be waged internally against the neo-conservative version of imperialism as well as against the continuation of neo-liberalism at the economic level. The class power ranged behind neo-liberalism, for example, is formidable, but the more problematic the neo-conservative form of governance appears, both internally and internationally, the more there will likely be division and dissent even within the elite classes over the direction the territorial logic of power should take. The current difficulties within the neo-liberal model and the threat it now poses to the United States itself may even provoke calls for an alternative logic of territorial power to be constructed. Whether or not that happens depends critically upon the balance of political forces within the United States. While this may not be determinant it will play a huge role in our individual and collective futures. With respect to that the rest of the world can only watch, wait, and hope. But one certain thing can be said. Across-the-board anti-Americanism from the rest of the world will not and cannot help. Those struggling in the United States to construct an alternative, both internally and with respect to foreign engagements, need all the sympathy and support they can get. In the same way that the inner/outer dialectic plays such a crucial role in the construction of neo-conservative imperialism, so a reversal of that dialectic has a crucial role to play in anti-imperialist politics.

Further Reading

S. Amin, *Imperialism and Unequal Development*, (New York: Monthly Review Press, 1977).

J. Atlas, 'A Classicist's Legacy: New Empire Builders,' *New York Times*, Week in Review, Sunday, 4 May 2003, pp. 1 and 4.

W. Bello, *Deglobalization: Ideas for a New World Economy* (London: Zed Books, 2002).

M. Boot, *The Savage Wars of Peace: Small Wars and the Rise of American Power* (New York: Basic Books, 2002).

K. Boulding and T. Mukerjee (eds.), *Economic Imperialism: A Book of Readings* (Ann Arbor: University of Michigan Press, 1972).

J. Cavanaugh, J. Mander, et al., *Alternatives to Globalization* (San Francisco: Bennett-Koehler, 2002).

J. Comaroff and J. Comaroff (eds.), *Millennial Capitalism and the Culture of Neoliberalism* (Durham, NC: Duke University Press, 2001).

R. Falk, *Predatory Globalization: A Critique* (Cambridge: Polity Press, 1999).

N. Ferguson, *Empire: The Rise and Demise of the British World Order and the Lessons of Global Power* (New York: Basic Books, 2003).

Further Reading

W. Finnegan, 'The Economics of Empire: Notes on the Washington Consensus,' *Harper's Magazine*, vol. 306, no. 1836 (May, 2003), pp. 41–54.

S. George and F. Sabelli, *Faith and Credit* (Harmondsworth: Penguin, 1995).

S. Hersh, 'Annals of National Security: How the Pentagon Outwitted the C.I.A.' *The New Yorker* (12 May 2003), pp. 44–51.

P. Hirst and G. Thompson, *Globalization in Question: The International Economy and the Possibility of Global Governance*, (Cambridge: Polity Press, rev. edn. 1999).

E. Hobsbawm, *The Age of Empire, 1875–1914* (London: Weidenfeld & Nicolson, 1987).

J. A. Hobson, *Imperialism* (Ann Arbor: University of Michigan Press, ed. with a new introduction by P. Siegelman, 1965 edn.).

D. Judd, *Radical Joe: A Life of Joseph Chamberlain* (London: Hamish Hamilton, 1977).

R. Kagan, *Of Paradise and Power: America and Europe in the New World Order* (New York: Knopf, 2003).

V. Kiernan, *America: The New Imperialism* (London: Zed Books, 1978).

N. Klein, *No Logo* (New York: Picador, 2000).

D. Korton, *When Corporations Rule the World* (Bloomfield, CT: Kummarian Press, 2001).

C. Kupchan, *The End of the American Era in US Foreign Policy and the Geopolitics of the 21st Century* (New York: Knopf, 2002).

B. Lewis, *What Went Wrong: Western Impact and Middle Eastern Response* (Oxford: Oxford University Press, 2001).

H. Mackinder, *Democratic Ideals and Reality* (New York: Norton, ed. by A.J. Pearce, 1962 edn).

H. Magdoff, *The Age of Imperialism: The Economics of U.S. Foreign Policy* (New York: Monthly Review Press, 1969).

Further Reading

K. E. Meyer, *The Dust of Empire: The Race for Mastery in the Asian Heartland* (New York: Public Affairs, 2003).

M. Mies, *Patriarchy and Accumulation on a World Scale: Women in the International Division of Labor* (London: Zed Books, 1999 edn.).

W. Nordhaus, 'Iraq: The Economic Consequences of War,' *The New York Review of Books*, vol. xlix, no. 19 (5 December 2002), pp. 9–12.

R. Owen and B. Sutcliffe (eds.), *Studies in the Theory of Imperialism* (London: Longman, 1972).

Oxfam International, *Rigged Rules and Double Standards* (London: Oxfam International, 2002).

H. Radice, *International Firms and Modern Imperialism* (Harmondsworth: Penguin, 1975).

S. Sassen, *Globalization and Its Discontents* (New York: New Press, 1998).

A. Schlesinger, *The Cycles of American History* (Boston: Houghton Mifflin, 1980).

V. Shiva, *Biopiracy: The Plunder of Nature and Knowledge* (Boston: South End Press, 1997).

V. Shiva, *Protect or Plunder? Understanding Intellectual Property Rights* (London: Zed Books, 2001).

V. Shiva, *Water Wars: Privatization, Pollution and Profit* (London: Zed Books, 2002).

K. Singh, *The Globalisation of Finance: A Citizen's Guide* (London: Zed Books, 1999).

G. Soros, *George Soros on Globalization* (New York: Public Affairs, 2002).

R. Steven, *Japan's New Imperialism* (Armonk, NY: M. E. Sharpe, 1990).

J. Stiglitz, *Globalization and Its Discontents* (New York: Norton, 2002).

A. Thornton, *Doctrines of Imperialism* (New York: Wiley, 1965).

Further Reading

A. K. Weinberg, *Manifest Destiny* (Baltimore: Johns Hopkins University Press, 1935).

D. Yergin, *The Prize: The Epic Quest for Oil, Money and Power* (New York: Simon and Schuster, 1991).

Bibliography

Press Comment

Altman, D., 'China: Partner, Rival or Both', *New York Times*, 2 Mar. 2003, Money and Business section, pp. 1 and 11.

Banerjee, N., 'Energy Companies Weigh their Possible Future in Iraq', *New York Times*, 26 Oct. 2002, p. C3.

Bush, G. W., 'Securing Freedom's Triumph', *New York Times*, 11 Sept. 2002, p. A33.

Cooper, R., 'The New Liberal Imperialism', *Observer*, 7 Apr. 2002.

Crampton, T., 'A Strong China May Give Boost to its Neighbors', *International Herald Tribune*, Economic Outlook, 23 Jan. 2003, pp. 16–17.

de Acule, C., 'Keeping a Wary Eye on the Housing Boom', *International Herald Tribune*, 23 Jan. 2003, p. 11.

Eckholm, E., 'Where Workers, Too, Rust, Bitterness Boils Over', *New York Times*, 20 Mar. 2002, p. A4.

Editorial, *Buenos Aires Herald*, 31 Dec. 2002, p. 4.

Fisk, R., 'The Case Against War: A Conflict Driven by the Self-Interest of America', *Independent*, 15 Feb. 2003, p. 20.

—— 'This Looming War isn't about Chemical Warheads or Human Rights: It's about Oil', *Independent*, 18 Jan. 2003, p. 18.

Bibliography

Friedman, T., 'A War for Oil?', *New York Times*, 5 Jan. 2003, Week in Review section, p. 11.

Hilterman, J., 'Halabja: America Didn't Seem to Mind Poison Gas', *International Herald Tribune*, 17 Jan. 2003, p. 8.

Ignatieff, M., 'The Burden', *New York Times*, 5 Jan. 2003, Sunday Magazine, pp. 22–54, repr. as 'Empire Lite', in *Prospect* (Feb. 2003), 36–43.

—— 'How to Keep Afghanistan from Falling Apart: The Case for a Committed American Imperialism', *New York Times*, 26 July 2002, Sunday Magazine, pp. 26–58.

Kahn, J., 'China Gambles on Big Projects for its Stability', *New York Times*, 13 Jan. 2003, pp. A1 and A8.

—— 'Made in China, Bought in China', *New York Times*, 5 Jan. 2003, Business section, pp. 1 and 10.

Kirkpatrick, D., 'Mr Murdoch's War', *New York Times*, 7 Apr. 2003, p. C1.

Krueger, A., 'Economic Scene', *New York Times*, 3 Apr. 2003, p. C2.

Madrick, J., 'The Iraqi Time Bomb', *New York Times*, 6 Apr. 2003, Sunday Magazine, p. 48.

Rosenthal, E., 'Workers' Plight Brings New Militancy to China', *New York Times*, 10 Mar. 2003, p. A8.

Tyler, P., 'Threats and Responses. News Analysis: A Deepening Fissure', *New York Times*, 6 Mar. 2003, p. 1.

Books and Journal Articles

Amin, S., 'Imperialism and Globalization', *Monthly Review* (June 2001), 1–10.

—— 'Social Movements at the Periphery', in P. Wignaraja (ed.), *New Social Movements in the South: Empowering the People* (London: Zed Books, 1993), 76–100.

Bibliography

Anderson, J., 'American Hegemony after September 11: Allies, Rivals and Contradictions', unpublished manuscript, Centre for International Borders Research, Queen's University, Belfast, 2002.

Anderson, P., 'Internationalism: A Breviary', *New Left Review*, 14 (Mar. 2002), 20.

Arendt, H., *Imperialism* (New York: Harcourt Brace Janovich, 1968 edn.).

Armstrong, D., 'Dick Cheney's Song of America: Drafting a Plan for Global Dominance', *Harper's Magazine*, 305 (Oct. 2002), 76–83.

Armstrong, P., A. Glyn, and J. Harrison, *Capitalism since World War II: The Making and Break Up of the Great Boom* (Oxford: Basil Blackwell, 1991).

Arrighi, G., *The Long Twentieth Century: Money, Power, and the Origins of our Times* (London: Verso, 1994).

—— and B. Silver, *Chaos and Governance in the Modern World System* (Minneapolis: University of Minnesota Press, 1999).

Baran, P., and P. Sweezy, *Monopoly Capital: An Essay on the American Economic and Social Order* (New York: Monthly Review Press, 1966).

Berman, M., 'Justice/Just Us: Rap and Social Justice in America', in A. Merrifield and E. Swyngedouw (eds.), *The Urbanization of Injustice* (New York: New York University Press, 1997), 161–79.

Bhagwati, J., 'The Capital Myth: The Difference between Trade in Widgets and Dollars', *Foreign Affairs*, 77/3 (1998), 7–12.

Bleaney, M., *Underconsumption Theories* (London: Methuen, 1976).

Blum, W., *Rogue State: A Guide to the World's Only Superpower* (London: Zed Books, 2002).

Bowden, B., 'Reinventing Imperialism in the Wake of September 11', *Alternatives: Turkish Journal of International*

Bibliography

Relations, 1/2 (Summer 2002); online at <http://alternatives.journal.fatih.edu.tr/Bowden.htm>.

Brenner, R., *The Boom and the Bubble: The U.S. in the World Economy* (London: Verso, 2002).

Brewer, A., *Marxist Theories of Imperialism* (London: Routledge & Kegan Paul, 1980).

Burkett, P., and M. Hart-Landsberg, 'Crisis and Recovery in East Asia: The Limits of Capitalist Development', *Historical Materialism*, 8 (2001), 3–48.

Cain, P., *Hobson and Imperialism: Radicalism, New Liberalism and Finance, 1887–1938* (Oxford: Oxford University Press, 2003).

Carchedi, G., 'Imperialism, Dollarization and the Euro', in Leo Panitch and Colin Leys (eds.), *Socialist Register 2002* (London: Merlin Press, 2001), 153–74.

Chamberlain, E., *The Theory of Monopolistic Competition* (Cambridge, Mass.: Harvard University Press, 1933).

Chomsky, N., *9-11* (New York: Seven Stories Press, 2001).

Doyle, M. W., *Empires* (Ithaca, NY: Cornell University Press, 1986).

Freeman, C., *High Tech and High Heels in the Global Economy* (Durham, NC: Duke University Press, 2000).

Gills, B. (ed.), *Globalization and the Politics of Resistance* (New York: Palgrave, 2001).

Gowan, P., *The Global Gamble: Washington's Faustian Bid for World Dominance* (London: Verso, 1999).

—— L. Panitch, and M. Shaw, 'The State, Globalization and the New Imperialism: A Round Table Discussion', *Historical Materialism*, 9 (2001), 3–38.

Guilbaut, S., *How New York Stole the Idea of Modern Art*, trans. A. Goldhammer (Chicago: University of Chicago Press, 1985 edn.).

Hardt, M., and A. Negri, *Empire* (Cambridge, Mass.: Harvard University Press, 2000).

Bibliography

Hart, G., *Disabling Globalization: Places of Power in Post-Apartheid South Africa* (Berkeley: University of California Press, 2002).

Harvey, D., *The Condition of Postmodernity* (Oxford: Basil Blackwell, 1989).

——— *The Limits to Capital* (Oxford: Basil Blackwell, 1982; repr. London: Verso Press, 1999).

——— *Paris, the Capital of Modernity* (New York: Routledge, 2003).

——— *Spaces of Capital: Towards a Critical Geography* (New York: Routledge, 2001).

——— *Spaces of Hope* (Edinburgh: Edinburgh University Press, 2000).

——— *The Urban Experience* (Baltimore: Johns Hopkins University Press, 1989).

Hegel, G. W., *The Philosophy of Right*, trans. T. Knox (New York: Oxford University Press, 1967 edn.).

Henderson, J., 'Uneven Crises: Institutional Foundations of East Asian Economic Turmoil', *Economy and Society*, 28/3 (1999), 327–68.

Hill, C., *The World Turned Upside Down* (Harmondsworth: Penguin, 1984).

Hines, C., *Localization: A Global Manifesto* (London: Earthscan, 2000).

Historical Materialism, 8 (2001), special issue: 'Focus on East Asia after the Crisis'.

Hofstadter, R., *The Paranoid Style in American Politics and Other Essays* (Cambridge, Mass.: Harvard University Press, 1996 edn.).

Huntington, S., *The Clash of Civilizations and the Remaking of the World Order* (New York: Simon & Schuster, 1997).

Isard, W., *Location and the Space Economy* (Cambridge, Mass.: MIT Press, 1956).

Johnson, C., *Blowback: The Costs and Consequences of American Empire* (New York: Henry Holt, 2000).

221

Bibliography

Julien, C.-A., J. Bruhat, C. Bourgin, M. Crouzet, and P. Renouvin, *Les Politiques d'expansion impérialiste* (Paris: Presses Universitaires de France, 1949).

Kennedy, P., *The Rise and Fall of the Great Powers: Economic Change and Military Conflict from 1500 to 2000* (New York: Fontana Press, 1990).

Klare, M., *Resource Wars: The New Landscape of Global Conflict* (New York: Henry Holt, 2001).

Krugman, P., *Development, Geography and Economic Theory* (Cambridge, Mass.: MIT Press, 1995).

Lee, C. K., *Gender and the South China Miracle: Two Worlds of Factory Women* (Berkeley: University of California Press, 1998).

Lefebvre, H., *The Survival of Capitalism: Reproduction of the Relations of Production*, trans. F. Bryant (New York: St Martin's Press, 1976).

Lenin, V. I., 'Imperialism: The Highest Stage of Capitalism', in *Selected Works*, vol. i (Moscow: Progress Publishers, 1963).

Li, S.-M., and W.-S. Tang, *China's Regions, Polity and Economy: A Study of Spatial Transformation in the Post-Reform Era* (Hong Kong: Chinese University Press, 2000).

Lösch, A., *The Economics of Location*, trans. William H. Woglom with the assistance of Wolfgang F. Stolper (New Haven: Yale University Press, 1954).

Luxemburg, R., *The Accumulation of Capital*, trans. A Schwarzschild (New York: Monthly Review Press, 1968 edn.).

McDonald, D., and J. Pape, *Cost Recovery and the Crisis of Service Delivery in South Africa* (London: Zed Books, 2002).

Markusen, A., *Profit Cycles, Oligopoly and Regional Development* (Cambridge, Mass.: MIT Press, 1985).

—— *Regions: The Economics and Politics of Territory* (Totowa, NJ: Rowman & Littlefield, 1987).

Marx, K., *Capital*, trans. B. Fowkes (New York: Viking, 1976).

—— and F. Engels, *On Colonialism* (New York: International Publishers, 1972).

Bibliography

Mehta, U., *Liberalism and Empire* (Chicago: University of Chicago Press, 1999).

Mittelman, J., *The Globalization Syndrome: Transformation and Resistance* (Princeton: Princeton University Press, 2000).

Morton, A., 'Mexico, Neoliberal Restructuring and the EZLN: A Neo-Gramscian Analysis', in B. Gills (ed.), *Globalization and the Politics of Resistance* (New York: Palgrave, 2001), 255–79.

Nash, J., *Mayan Visions: The Quest for Autonomy in an Age of Globalization* (New York: Routledge, 2001).

National Security Strategy of the United State of America at <www.whitehouse.gov/nsc/nss>.

Nye, J., *The Paradox of American Power: Why the World's Only Super-Power Cannot Go It Alone* (Oxford: Oxford University Press, 2003).

Ong, A., *Spirits of Resistance and Capitalist Discipline: Factory Women in Malaysia* (Albany: State University of New York Press, 1987).

Panitch, L., 'The New Imperial State', *New Left Review*, 11/1 (2000), 5–20.

Payer, C., *The Debt Trap: The IMF and the Third World* (New York: Monthly Review Press, 1974).

Perelman, M., *The Invention of Capitalism: Classical Political Economy and the Secret History of Primitive Accumulation* (Durham, NC: Duke University Press, 2000).

Petras, J., and H. Veltmeyer, *Globalization Unmasked: Imperialism in the 21st Century* (London: Zed Books, 2001).

Pilger, J., *The New Rulers of the World* (London: Verso, 2002).

Pollard, S., *Essays on the Industrial Revolution in Britain*, ed. Colin Holmes (Aldershot: Ashgate Variorum, 2000).

Rostow, W. W., *The Stages of Economic Growth: A Non-Communist Manifesto* (Cambridge: Cambridge University Press, 1966 edn.).

Roy, A., *Power Politics* (Cambridge, Mass.: South End Press, 2001).

Bibliography

Servan-Schreiber, J. J., *American Challenge* (New York: Scribner, 1968).

Smith, N., *American Empire: Roosevelt's Geographer and the Prelude to Globalization* (Berkeley: University of California Press, 2003).

Soederberg, S., 'The New International Financial Architecture: Imposed Leadership and "Emerging Markets"', in Leo Panitch and Colin Leys (eds.), *Socialist Register 2002* (London: Merlin Press, 2001), 175–92.

Strange, S., *Mad Money: When Markets Outgrow Governments* (Ann Arbor: University of Michigan Press, 1998).

Thompson, E. P., *The Making of the English Working Class* (Harmondsworth: Penguin, 1968).

'U.S. Imperial Ambitions and Iraq' [editorial], *Monthly Review*, 54/7 (2002), 1–13.

Wade, R., and F. Veneroso, 'The Asian Crisis: The High Debt Model versus the Wall Street–Treasury–IMF Complex', *New Left Review*, 228 (1998), 3–23.

Warren, B., *Imperialism: Pioneer of Capitalism* (London: Verso, 1981).

Went, R., 'Globalization in the Perspective of Imperialism', *Science and Society*, 66/4 (2002–3), 473–97.

Williams, W. A., *Empire as a Way of Life* (New York: Oxford University Press, 1980).

Wolf, E., *Peasant Wars of the Twentieth Century* (New York: HarperCollins, 1969).

Yergin, D., J. Stanislaw, and D. Tergin, *The Commanding Heights: The Battle Between Government and Market Place that is Remaking the Modern World* (New York: Simon & Schuster, 1999).

Zhang, L., *Strangers in the City: Reconfigurations of Space, Power and Social Networks within China's Floating Population* (Stanford: Stanford University Press, 2001).

Notes

1. All About Oil

1. M. Berman, 'Justice/Just Us: Rap and Social Justice in America', in A. Merrifield and E. Swyngedouw (eds.), *The Urbanization of Injustice* (New York: New York University Press, 1997), 148.
2. M. Ignatieff, 'The Burden', *New York Times*, 5 Jan. 2003, Sunday Magazine, pp. 22–54, repr. as 'Empire Lite', in *Prospect* (Feb. 2003), 36–43. See also id., 'How to Keep Afghanistan from Falling Apart: The Case for a Committed American Imperialism', *New York Times*, 26 July 2002, Sunday Magazine, pp. 26–58.
3. Many of these quotations are conveniently collected together in B. Bowden, 'Reinventing Imperialism in the Wake of September 11', *Alternatives: Turkish Journal of International Relations*, 1/2 (Summer 2002). This can be found online at <http://alternatives.journal.fatih.edu.tr/Bowden.htm>.
4. G. W. Bush, 'Securing Freedom's Triumph', *New York Times*, 11 Sept. 2002, p. A33. *The National Security Strategy of the United State of America* can be found on the website <www.whitehouse.gov/nsc/nss>. Ignatieff, 'The Burden',

opens his argument (p. 22) with a discussion of Bush's West Point speech.

5. M. W. Doyle, *Empires* (Ithaca, NY: Cornell University Press, 1986), provides an interesting comparative study of empires. For the US case see also W. A. Williams, *Empire as a Way of Life* (New York: Oxford University Press, 1980).

6. The topic of the 'new imperialism' has been broached on the left in L. Panitch, 'The New Imperial State', *New Left Review*, 11/1 (2000), 5–20; see also P. Gowan, L. Panitch, and M. Shaw, 'The State, Globalization and the New Imperialism: A Round Table Discussion', *Historical Materialism*, 9 (2001), 3–38. Other commentaries of interest are J. Petras and H. Veltmeyer, *Globalization Unmasked: Imperialism in the 21st Century* (London: Zed Books, 2001); R. Went, 'Globalization in the Perspective of Imperialism', *Science and Society*, 66/4 (2002–3), 473–97; S. Amin, 'Imperialism and Globalization', *Monthly Review* (June 2001), 1–10; and M. Hardt and A. Negri, *Empire* (Cambridge, Mass.: Harvard University Press, 2000).

7. Cited in C. Johnson, *Blowback: The Costs and Consequences of American Empire* (New York: Henry Holt, 2000), 18.

8. J. Hilterman, 'Halabja: America Didn't Seem to Mind Poison Gas', *International Herald Tribune*, 17 Jan. 2003, p. 8.

9. Reported in R. Fisk, 'The Case Against War: A Conflict Driven by the Self-Interest of America', *Independent*, 15 Feb. 2003, p. 20.

10. Reported in R. Fisk, 'This Looming War isn't about Chemical Warheads or Human Rights: It's about Oil', *Independent*, 18 Jan. 2003, p. 18. See also the website <www.newamericancentury.org>.

Notes

11. H. Arendt, *Imperialism* (New York: Harcourt Brace Janovich, 1968 edn.), 22.
12. N. Banerjee, 'Energy Companies Weigh their Possible Future in Iraq', *New York Times*, 26 Oct. 2002, p. C3.
13. M. Klare, *Resource Wars: The New Landscape of Global Conflict* (New York: Henry Holt, 2001), provides an excellent overview of the geopolitics of oil.
14. The Editors, 'U.S. Imperial Ambitions and Iraq', *Monthly Review*, 54/7 (2002), 1–13.
15. J. Nye, *The Paradox of American Power: Why the World's Only Super-Power Cannot Go It Alone* (Oxford: Oxford University Press, 2003).
16. T. Friedman, 'A War for Oil?', *New York Times*, 5 Jan. 2003, Week in Review section, p. 11.

2. How America's Power Grew

1. G. Arrighi, *The Long Twentieth Century: Money, Power, and the Origins of our Times* (London: Verso, 1994), 33–4.
2. Arendt, *Imperialism*, 23.
3. Arrighi, *The Long Twentieth Century*, 62.
4. P. Kennedy, *The Rise and Fall of the Great Powers: Economic Change and Military Conflict from 1500 to 2000* (New York: Fontana Press, 1990).
5. G. Arrighi and B. Silver, *Chaos and Governance in the Modern World System* (Minneapolis: University of Minnesota Press, 1999), 26–8.
6. J. Mittelman, *The Globalization Syndrome: Transformation and Resistance* (Princeton: Princeton University Press, 2000), part II in particular; Mittelman is one of many

authors who have taken up the regionalization thesis seriously.

7. Johnson, *Blowback*; J. Pilger, *The New Rulers of the World* (London: Verso, 2002): W. Blum, *Rogue State: A Guide to the World's Only Superpower* (London: Zed Books, 2002); and, of course, N. Chomsky, *9-11* (New York: Seven Stories Press, 2001).

8. Arendt, *Imperialism*, 18.

9. Ibid. 32.

10. N. Smith, *American Empire: Roosevelt's Geographer and the Prelude to Globalization* (Berkeley: University of California Press, 2003).

11. R. Hofstadter, *The Paranoid Style in American Politics and Other Essays* (Cambridge, Mass.: Harvard University Press, 1996 edn.).

12. Smith, *American Empire*, 20.

13. The Editors, 'U.S. Imperial Ambitions and Iraq', 3–13.

14. This is a central thesis in the work of W. A. Williams, *Empire as a Way of Life*.

15. W. W. Rostow, *The Stages of Economic Growth: A Non-Communist Manifesto* (Cambridge: Cambridge University Press, 1966 edn.).

16. S. Guilbaut, *How New York Stole the Idea of Modern Art*, trans. A. Goldhammer (Chicago: University of Chicago Press, 1985 edn.).

17. J. J. Servan-Schreiber, *American Challenge* (New York: Scribner, 1968).

18. P. Armstrong, A. Glyn, and J. Harrison, *Capitalism since World War II: The Making and Break Up of the Great Boom* (Oxford: Basil Blackwell, 1991).

Notes

19. The significance of seigniorage is taken up by G. Carchedi in 'Imperialism, Dollarization and the Euro', in Leo Panitch and Colin Leys (eds.), *Socialist Register 2002* (London: Merlin Press, 2001), 153–74.

20. P. Gowan, *The Global Gamble: Washington's Faustian Bid for World Dominance* (London: Verso, 1999).

21. Ibid. 49, on frequency of debt crises.

22. P. Anderson, 'Internationalism: A Breviary', *New Left Review*, 14 (Mar. 2002), 20, notes how 'something like Kautsky's vision' had come to pass and that liberal theorists, like Robert Keohane, had already noted the connection.

23. R. Brenner, *The Boom and the Bubble: The U.S. in the World Economy* (London: Verso, 2002), 3.

24. Arrighi and Silver, *Chaos and Governance*, 31–3.

25. Gowan, *The Global Gamble*, 123.

26. S. Soederberg, 'The New International Financial Architecture: Imposed Leadership and "Emerging Markets"', in Panitch and Leys (eds.), *Socialist Register 2002*, 175–92.

27. Arrighi and Silver, *Chaos and Governance*, 288–9.

28. D. Armstrong, 'Dick Cheney's Song of America: Drafting a Plan for Global Dominance', *Harper's Magazine*, 305 (Oct. 2002), 76–83.

29. Mittelman, *The Globalization Syndrome*.

30. Cited in P. Tyler, 'Threats and Responses. News Analysis: A Deepening Fissure', *New York Times*, 6 Mar. 2003, p. 1.

Notes

3. Capital Bondage

1. H. Lefebvre, *The Survival of Capitalism: Reproduction of the Relations of Production*, trans. F. Bryant (New York: St Martin's Press, 1976).

2. Most of these essays from the 1970s and 1980s have been republished in D. Harvey, *Spaces of Capital: Towards a Critical Geography* (New York: Routledge, 2001). The main line of argument can also be found in D. Harvey, *The Limits to Capital* (Oxford: Basil Blackwell, 1982; repr. London: Verso Press, 1999).

3. My own version of this theoretical argument is detailed in Harvey, *Limits to Capital*, chs. 6 and 7.

4. A fascinating case of this is to be found in L. Zhang, *Strangers in the City: Reconfigurations of Space, Power and Social Networks within China's Floating Population* (Stanford: Stanford University Press, 2001).

5. Arrighi and Silver, *Chaos and Governance*, 48–9.

6. Harvey, *Limits to Capital*; id., *Spaces of Capital*; id., *The Urban Experience* (Baltimore: Johns Hopkins University Press, 1989).

7. W. Isard, *Location and the Space Economy* (Cambridge, Mass.: MIT Press, 1956).

8. E. Chamberlain, *The Theory of Monopolistic Competition* (Cambridge, Mass.: Harvard University Press, 1933); A. Lösch, *The Economics of Location*, trans. William H. Woglom with the assistance of Wolfgang F. Stolper (New Haven: Yale University Press, 1954).

9. P. Baran and P. Sweezy, *Monopoly Capital; An Essay on the American Economic and Social Order* (New York: Monthly Review Press, 1966).

Notes

10. See Harvey, *Limits to Capital*; id., *The Condition of Postmodernity* (Oxford: Basil Blackwell, 1989), pt. III.

11. P. Krugman, *Development, Geography and Economic Theory* (Cambridge, Mass.: MIT Press, 1995).

12. Mittelman, *The Globalization Syndrome*.

13. Harvey, *The Urban Experience*, ch. 5.

14. S. Pollard, *Essays on the Industrial Revolution in Britain*, ed. Colin Holmes (Aldershot: Ashgate Variorum, 2000), 219–71.

15. A. Markusen, *Regions: The Economics and Politics of Territory* (Totowa, NJ: Rowman & Littlefield, 1987); id., *Profit Cycles, Oligopoly and Regional Development* (Cambridge, Mass.: MIT Press, 1985); and S.-M. Li and W.-S. Tang, *China's Regions, Polity and Economy: A Study of Spatial Transformation in the Post-Reform Era* (Hong Kong: Chinese University Press, 2000).

16. D. Harvey, *Paris, the Capital of Modernity* (New York: Routledge, 2003).

17. Brenner, *The Boom and the Bubble*.

18. C. de Acule, 'Keeping a Wary Eye on the Housing Boom', *International Herald Tribune*, 23 Jan. 2003, p. 11.

19. I elaborate on Marx's category of 'fictitious capital' in Harvey, *Limits to Capital*, ch. 10. See also S. Strange, *Mad Money: When Markets Outgrow Governments* (Ann Arbor: University of Michigan Press, 1998).

20. Harvey, *Paris, the Capital of Modernity*.

21. C. Payer, *The Debt Trap: The IMF and the Third World* (New York: Monthly Review Press, 1974).

22. J. Kahn, 'China Gambles on Big Projects for its Stability', *New York Times*, 13 Jan. 2003, pp. A1 and A8; 'Made in

Notes

China, Bought in China', *New York Times*, 5 Jan. 2003, Business section, pp. 1 and 10; D. Altman, 'China: Partner, Rival or Both', *New York Times*, 2 Mar. 2003, Money and Business section, pp. 1 and 11; T. Crampton, 'A Strong China May Give Boost to its Neighbors', *International Herald Tribune*, Economic Outlook, 23 Jan. 2003, pp. 16–17.

23. G. W. Hegel, *The Philosophy of Right*, trans. T. Knox (New York: Oxford University Press, 1967 edn.).

24. V. I. Lenin, 'Imperialism: The Highest Stage of Capitalism', in *Selected Works*, vol. i (Moscow: Progress Publishers, 1963).

25. This whole common history of a radical shift from internal to external solutions to political-economic problems in response to the dynamics of class struggle across many capitalist states is told in a little-known but quite fascinating collection by C.-A. Julien, J. Bruhat, C. Bourgin, M. Crouzet, and P. Renouvin, *Les Politiques d'expansion impérialiste* (Paris: Presses Universitaires de France, 1949), in which the cases of Ferry, Chamberlain, Roosevelt, Crispi, and others are all examined in comparative detail.

26. P. Cain, *Hobson and Imperialism: Radicalism, New Liberalism and Finance, 1887–1938* (Oxford: Oxford University Press, 2003).

27. Arendt, *Imperialism*, 32.

28. J. Henderson, 'Uneven Crises: Institutional Foundations of East Asian Economic Turmoil', *Economy and Society*, 28/3 (1999), 327–68.

29. Gowan, *The Global Gamble*, 21.

30. Ibid., ch. 4.

31. Editorial, *Buenos Aires Herald*, 31 Dec. 2002, p. 4.

32. J. Bhagwati, 'The Capital Myth: The Difference between Trade in Widgets and Dollars', *Foreign Affairs*, 77/3 (1998), 7–12.

4. Accumulation by Dispossession

1. R. Luxemburg, *The Accumulation of Capital*, trans. A Schwarzschild (New York: Monthly Review Press, 1968 edn.).

2. See e.g. M. Bleaney, *Underconsumption Theories* (London: Methuen, 1976); A. Brewer, *Marxist Theories of Imperialism* (London: Routledge & Kegan Paul, 1980).

3. K. Marx, *Capital*, trans. B. Fowkes (New York: Viking, 1976), vol. i, ch. 25.

4. Arendt, *Imperialism*, 15, 28.

5. M. Perelman, *The Invention of Capitalism: Classical Political Economy and the Secret History of Primitive Accumulation* (Durham, NC: Duke University Press, 2000). There is also an extensive debate in *The Commoner* (<www. thecommoner.org>) on the new enclosures and on whether primitive accumulation should be understood as a purely historical or a continuing process. DeAngelis <http:// homepages.uel.ac.uk/M.DeAngelis> provides a good summary.

6. Marx, *Capital*, vol. i, pt. 8.

7. E. P. Thompson, *The Making of the English Working Class* (Harmondsworth: Penguin, 1968).

8. Contemporary ethnographies of proletarianization, many of which emphasize the importance of gender issues, illustrate some of the diversity very well. See e.g. A. Ong, *Spirits of Resistance and Capitalist Discipline: Factory Women in*

Notes

Malaysia (Albany: State University of New York Press, 1987); C. Freeman, *High Tech and High Heels in the Global Economy* (Durham, NC: Duke University Press, 2000); C. K. Lee, *Gender and the South China Miracle: Two Worlds of Factory Women* (Berkeley: University of California Press, 1998).

9. D. Kirkpatrick, 'Mr Murdoch's War', *New York Times*, 7 Apr. 2003, p. C1.

10. R. Wade and F. Veneroso, 'The Asian Crisis: The High Debt Model versus the Wall Street–Treasury–IMF Complex', *New Left Review*, 228 (1998), 3–23.

11. Ibid. Other accounts of this crisis are given in Henderson, 'Uneven Crises'; Johnson, *Blowback*, ch. 9; and the special issue of *Historical Materialism*, 8 (2001), 'Focus on East Asia after the Crisis', particularly P. Burkett and M. Hart-Landsberg, 'Crisis and Recovery in East Asia: The Limits of Capitalist Development', pp. 3–48.

12. Gowan, *The Global Gamble*, offers a compelling account.

13. E. Eckholm, 'Where Workers, Too, Rust, Bitterness Boils Over', *New York Times*, 20 Mar. 2002, p. A4; E. Rosenthal, 'Workers' Plight Brings New Militancy to China', *New York Times*, 10 Mar. 2003, p. A8.

14. D. Yergin, J. Stanislaw, and D. Tergin, *The Commanding Heights: The Battle Between Government and Market Place that is Remaking the Modern World* (New York: Simon & Schuster, 1999).

15. D. McDonald and J. Pape, *Cost Recovery and the Crisis of Service Delivery in South Africa* (London: Zed Books, 2002), 162.

16. J. Nash, *Mayan Visions: The Quest for Autonomy in an Age of Globalization* (New York: Routledge, 2001), 81–4.

Notes

17. A. Roy, *Power Politics* (Cambridge, Mass.: South End Press, 2001), 16.
18. Ibid. 43.
19. C. Hill, *The World Turned Upside Down* (Harmondsworth: Penguin, 1984).
20. For Marx on India see the collection K. Marx and F. Engels, *On Colonialism* (New York: International Publishers, 1972); B. Warren, *Imperialism: Pioneer of Capitalism* (London: Verso, 1981).
21. B. Gills (ed.), *Globalization and the Politics of Resistance* (New York: Palgrave, 2001), is an excellent collection reflecting some of this diversity.
22. Roy, *Power Politics*, 86.
23. Nash, *Mayan Visions*; A. Morton, 'Mexico, Neoliberal Restructuring and the EZLN: A Neo-Gramscian Analysis', in Gills (ed.), *Globalization*, 255–79.
24. Hardt and Negri, *Empire*.
25. S. Amin, 'Social Movements at the Periphery', in P. Wignaraja (ed.), *New Social Movements in the South: Empowering the People* (London: Zed Books, 1993), 95. This collection has several pieces that reflect perceptively on these themes.
26. E. Wolf, *Peasant Wars of the Twentieth Century* (New York: HarperCollins, 1969); Nash, *Mayan Visions*; Morton, 'Mexico'.
27. A particularly strong version of this argument is given in C. Hines, *Localization: A Global Manifesto* (London: Earthscan, 2000). See also Wignaraja (ed.), *New Social Movements*.
28. Roy, *Power Politics*.
29. Arendt, *Imperialism*, 28.

Notes

5. Consent to Coercion

1. G. Hart, *Disabling Globalization: Places of Power in Post-Apartheid South Africa*, (Berkeley: University of California Press, 2002).
2. Klaus Schwab and Claude Smadja, cited in D. Harvey, *Spaces of Hope* (Edinburgh: Edinburgh University Press, 2000), 70.
3. S. Huntington, *The Clash of Civilizations and the Remaking of the World Order* (New York: Simon & Schuster, 1997).
4. The website is <www.newamericancentury.org>.
5. See Smith, *American Empire*.
6. This formula is well described in Johnson, *Blowback*.
7. Armstrong, 'Dick Cheney's Song of America'.
8. A. Krueger, 'Economic Scene', *New York Times*, 3 Apr. 2003, p. C2; J. Madrick, 'The Iraqi Time Bomb', *New York Times*, 6 Apr. 2003, Sunday Magazine, p. 48.
9. R. Cooper, 'The New Liberal Imperialism', *Observer*, 7 Apr. 2002. The critique mounted in U. Mehta, *Liberalism and Empire* (Chicago: University of Chicago Press, 1999), is simply devastating when put up against Cooper's formulations. I have on this point, as elsewhere, benefited greatly from the analysis laid out in J. Anderson, 'American Hegemony after September 11: Allies, Rivals and Contradictions', unpublished manuscript, Centre for International Borders Research, Queen's University, Belfast, 2002.

Index

237

Index

anti-capitalism 169–80
anti-globalization 74, 162–3, 188–9,
202
anti-imperialism 169–80
Arendt, Hannah 15–16
and accumulation by dispossession 140, 142–3, 182, 233
and capital bondage 127, 232
and consent to coercion 193,
196–7
and growth of US power 34–5,
44–5, 51, 227
Argentina 59, 117, 118, 120,
159–60, 206
Armageddon, belief in 191
Armitage, Richard 15
arms race *see* militarism
Armstrong, D. 80, 229, 236
Armstrong, P. 228
Arrighi, Giovanni viii
and capital bondage 93, 230
and growth of US power 27, 36,
41, 72, 77, 227, 229
and potential US domination
74–5
Ashcroft, John 17, 192
asymmetry in exchange 130
austerity, future 208–9
Australia vii, 7–8

balance of payments deterioration
71–2
Balkans 19, 85
Banerjee, N. 227
Bangladesh 134
Baran, P. 97, 230
Belgian empire 45
'benevolence' 40
Berman, Marshall 2–3, 225
Bhagwati, J. 131, 233
Bill of Rights (US) 38, 211
biological and chemical weapons
9–10, 13–14
biopiracy 148

Birmingham 104, 125
Bismarck, Prince Otto von 92
Blair, Tony 209
Bleaney, M. 233
Blum, W. 38, 228
Boer War 126, 180, 181
Boot, Max 4
bourgeois imperialism 34–5, 188,
211–12
rise of (1870–1945) 42–9
second stage in rule of (1945–70)
49–62, 68
society 81, 90, 125
see also colonialism
see also neo-liberalism
Bourgin, C. 232
Bowden, B. 225
Brandt, Willy 85
Braudel, F. 72
Brazil 176
and capital bondage 121, 130,
134, 135
and consent to coercion 186,
202
and growth of US power 53, 59
Brenner, R. 71, 108, 112, 121, 229,
231
Bretton Woods 54, 62, 155
Bright, John 104
Britain
and accumulation by dispossession 142, 162, 167, 182
privatization 149, 157–9
see also empire *below*
and capital bondage 104, 106,
112–13
inner contradictions 125–6
spatio-temporal fix 117, 119
state and 91–2, 132, 134
and consent to coercion 184, 186,
195, 196, 198, 200, 201
empire and decline of 1–3, 19, 20,
34, 35, 45, 46, 120
free trade 133

238

Index

and growth of US power 67, 79, 82

and Iraq invasion vii, 7–8, 11

labour militancy 63

and oil 18, 20, 23

and Second World War 49

Bruhat, J. 232

Bulgaria 199

Burkett, P. 234

Burns, William 10

Bush, George W.

aggressive foreign policy 6, 12–18, 75, 77, 83, 192, 225

towards Iraq 6, 8, 10–11, 14–15, 81, 93, 180

electoral appeal 17

electoral legitimacy questionable 13, 190

and Eurasian power 85

and free trade 133

oil industry links 18

and Pax Americana 4–5

Cain, P. 232

Canada 23, 90, 119

capital

circuits of 108–15

fixed *see* infrastructure; spatio-temporal fix

-labour pact 58

logic of power versus territorial 27–30, 33–6, 101–8, 183

see also accumulation; capital bondage

capital bondage 87–136

circuits of capital 108–15

inner contradictions 125–7

mediating institutions 127–32

political/territorial versus capitalist logics of power 101–8

space economy, production of 88–9, 94–101

state powers and capital accumulation 88, 89–93, 132–6

see also accumulation; spatio-temporal fix

capitalist imperialism as fusion of politics of state and empire 26–33, 46–7

and territory 33–6

see also finance; growth of US power; hegemony

Carchedi, G. 229

Carter, Jimmy 21, 180, 195

Caspian Basin oil reserves 24–5

and China 25, 78, 124, 196

Central America *see* Latin America

central Europe *see* eastern and central Europe

Chamberlain, E. 96, 230

Chamberlain, Joseph 104, 125–6, 135, 180

Chavez, Hugo 8–9, 24, 78

Cheney, Dick 18, 79, 192

Chiapas, Zapatista rebellion in 160, 161, 168, 174–6, 189

Chile 8, 38, 53, 56, 59, 172

China

and accumulation by dispossession capitalism 149, 153–4, 155, 156

collectivization 165

Opium Wars 138

and capital bondage 105, 113, 131

spatio-temporal fix 117, 120–4

state 90, 92, 134

and consent to coercion 187, 194, 196, 199, 200, 208–9

and growth of US power 49, 52–4 *passim*, 69, 71, 72, 77–8, 84

imperial 5

and Iraq invasion 8, 78, 200

and oil 18, 23, 25, 77–8, 124, 196, 209

and US imperialism 7, 8

Chomsky, Noam 38, 228

Index

Index

crises of capitalism 61, 121–2, 138,
150–6, 183, 184
see also debt
Crouzet, M. 232
crusade attitude 191
Cuba 38, 54, 166
culture, American *see*
Americanization

dams 123, 166, 177–8
Darwinism, social 45
de Gaulle, Charles 85, 92
DeAngelis, M. 233
debt 67, 77, 117–18, 147, 186
debtor nation, US as biggest 13
and Iraq invasion 205, 206
see also aid and loans
decolonization and self-government
2, 54, 58
de-industrialization 41, 64–5, 155,
185
democracy
as motive for Iraq invasion 9
crushed by US 9, 20, 38, 53–4
subverted 59–60
dependency theory 7
devaluation 39, 66, 70–1, 150, 155
dirigisme 92
dispossession *see* accumulation by
dispossession; colonialism
distance, friction of 94, 98–9
distributive power 37
domination *see* coercion; hegemony
Dominican Republic 56, 59
Doyle, M. W. 226
dual domains of anti-capitalism and
anti-imperialism 169–80
Dust Bowl 155–6

East Asia
and accumulation by dispossession 138, 149, 150–6 *passim*,
161, 165, 171
and capital bondage 92, 97, 105

circuits of capital 112, 113
mediating institutions, power
of 127–8, 131
spatio-temporal fix 117–18,
120–4
state 90, 92, 132, 134
and consent to coercion 185, 187,
194–5, 198–9, 203–4
and growth of US power 39, 41,
77, 83
neo-liberal hegemony 65, 66,
68–74 *passim*
see also China; Japan; North
Korea; South Korea; South-
East Asia
East India Company 90
eastern and central Europe 6, 50,
82, 85, 165, 199
Eckholm, E. 234
economy
boom 16
de-industrialization 41, 64–5,
155, 185
emulation 61–2
and growth of US power 48, 51,
54–7, 64–72
power misused 38–9
liquidations 38–9
military *see under* militarism
recession 13
see also accumulation; consumerism; globalization
education
investment in 111, 115, 125
school massacre 16
Egypt 203
Eisenhower, Dwight D. 2, 6, 60,
192
ejido lands privatized *see* Chiapas
empires *see* colonialism; imperialism
emulation 41–2, 61–2
Engels, F. 235
Enlightenment 99
Enron 13, 147, 178

241

Index

Index

Index

Index

Index

246

Index

247

Index

New York City (*cont.*)
  metropolitan imperialism 32
  World Trade Center *see* 9/11
New York Times
  on American empire 3
  on 9/11 4–5
  on torture 10
  on Venezuelan coup 8, 9
NGOs 167, 189
Nicaragua 165
NICs 155
  see also East Asia; South-East
    Asia
9/11 71, 100
  and consent to coercion 190,
    193–6 *passim*
  as justification for Iraq invasion
    15, 17, 193–4, 196
  New York Times on 4–5
  and recession 13
  seen as God's anger 190–4
    passim
  war more acceptable after 22–3
Nixon, Richard 6, 128
non-governmental organizations
  167, 189
North Korea 22, 79
  and consent to coercion 194, 195,
    198
  nuclear weapons 10
  and oil 25
North Sea oil 23
nuclear weapons 10–11
Nye, Joseph 21, 227

OECD 55, 59
Ogoni people 166
oil 18–25, 150
  and consent to coercion 195, 202,
    204–5
  and Iraq 2, 18–19, 22, 23, 24
  needed for Iraq's recovery 204–5
  OPEC 9, 21, 62, 128
  reserves unknown 23–4

  see also Middle East
Oklahoma 16
oligopoly 129–31
Olympia & York 114–15
Olympic Games 123
Ong, A. 233–4
OPEC 9, 21, 62, 128
Opium Wars 138
Organization for Economic
  Cooperation and
  Development 55, 59
Organization of Petroleum
  Exporting Countries 9, 21,
  62, 128
'original' accumulation *see* primitive
Osama bin Laden 14, 198
overaccumulation/overextension *see*
  accumulation
overreach problem 60–1

Pahlavi, Shah Muhammad Reza 9,
  21, 53
Pakistan 203
Palestinians 20, 191
Pan-Arabism, US fear of 14–15
Panitch, Leo 226, 229
Pape, J. 234
'paranoid style' of US politics 49
Paris
  Club 118
  rebuilt 43, 106, 114
Patriot Act 193
patronage 48, 54
Pax Americana *see* peace
Payer, Cheryl 118, 231
peace (Pax Americana) 4–6, 199
  as goal for going to war 5, 133
Pearl River delta 69, 105
pension funds 13, 71, 205
Perelman, M. 233
Perle, R. 15
Perroux, François 102
Petras, J. 226
physiocrats 92

248

Index

Index

Index

and consent to coercion 185–7, 206
and growth of US power 38, 39, 53, 59, 77, 83–4
neo-liberal hegemony 65–70 *passim*, 72, 73
see also Indonesia; Singapore; Taiwan; Thailand
Soviet Union, former 19, 49
and Afghanistan 52
as threat *see* Cold War
collapse 30–1, 149, 153
and eastern Europe *see* subordinate *below*
Second World War losses 49–50
subordinate client states 6, 50, 165, 199
see also Russia
space
economy, production of 88–9, 94–101
and time compressed 98
see also spatio-temporal fix
Spain, Iraq invasion and vii, 7–8, 82, 200
spatio-temporal fix 43–4
and accumulation by dispossession 139–40
and capital bondage 87–8, 89, 108–9, 115–24
stagflation 61
Stalin, Josef 49
Stanislaw, J. 234
state
and empire politics fused *see* capitalist imperialism
powers and capital accumulation 88, 89–93, 132–6
taxation 13, 91–2
terrorism 38
Steinbeck, John 155–6
Stern, Howard 16
Strange, S. 231
Straw, Jack 198

structural adjustment 66, 151–2, 184, 208–9
struggles
class *see under* class
overaccumulation by dispossession 160–1, 162–9, 174–6
see also wars
sub–imperialism 185–6
Suez Crisis (1956) 2
Suharto, Thojib 59, 171
Sukarno, Ahmed 171
Summers, Larry 6, 187, 192
surpluses of capital 88–9, 109
see also investment
surpluses of labour power 88–9, 109
surveillance 194
Sweezy, P. 97, 230
'switching crises' 121–2
Syria 19, 198, 200

Taiwan 187
and accumulation by dispossession 155, 156, 161
and capital bondage 92, 120, 121, 128, 130
and growth of US power 41, 53, 69, 72, 84
takeovers 67
Taliban 194, 196
Tang, W.-S. 231
taxation 13, 91–2
technology 98, 141
Tergin, D. 234
territorial power 93, 94
versus capitalist 27–30, 33–6, 101–8, 183
terrorism, US response to 4, 14, 38
see also 9/11 *and under* Afghanistan
Thailand 66
and capital bondage 112, 128, 135
and consent to coercion 186, 206
Thatcher, Margaret 157–8

Index

Index